QUIZ
SERIES

In the World of

SCIENCE

by
Jon Carlisle

Illustrations by
Tom Kerr
Editorial Cartoonist

All inquiries should be addressed to:

Barron's Educational Series, Inc.
250 Wireless Boulevard
Hauppauge, New York 11788

Library of Congress Catalog Card No. 92-23610

International Standard Book No. 0-8120-4854-7

Library of Congress Cataloging-in-Publication Data
Carlisle, Jon.
Who, what, when, where, why in the world of science / by Jon Carlisle ; illustrations by Tom Kerr.
p. cm. — (Barron's whiz quiz series)
ISBN 0-8120-4854-7
1. Science—Miscellanea. I. Title. II. Title: Science. III. Series.
Q173.C28 1992 92-23610
500—dc20 CIP

PRINTED IN THE UNITED STATES OF AMERICA
2550 987654321

Contents

To Stefan, Aaron, Zachary, Catherine, Jay, Daniel, Laura, Sky, and Jacqueline

Introduction

How do things work? Where did this world come from? How will it end? What makes rainbow light scatter off the wings of a butterfly?

And why do we wonder?

Fascination with these kinds of questions and the fact that you are reading this book points up a fact about you and all human beings: We are intensely curious creatures. We long to understand the patterns and processes in the world around us. You, and all your brothers and sisters on planet Earth, exhibit the uniquely human trait of *wanting to know*, a trait that motivates all scientific investigation.

The questions and answers in this book spring from the dedication and discoveries of men and women who have extended the frontiers of human knowledge by searching for truths about ourselves, our planet, our universe.

Every effort has been made by the author, and various experts in a number of scientific disciplines, to insure the accuracy of the material in this volume. But we must remember that what seems truth today will be discounted tomorrow. For the nature of science is to ever broaden our understanding and relentlessly re-examine what we think we know. That process continually synthesizes new explanations for the mysterious workings of the grand and wonderful cosmos in which we live.

Consider the history of what is called Newtonian Physics. Isaac Newton, in 1687, published his famous work, *Principia,* in which he summed up human knowledge of the laws of gravity and motion. For more than two hundred years, scientists believed that Newton had provided final answers to many of science's most puzzling questions. But then, in 1905, Albert Einstein, with his Theory of Relativity, called into question the laws spelled out by Newton. This new theory said that Newton's laws were true only under certain circumstances or were only approximations of the truth.

Someday Einstein's theories, and those of other modern-day physicists, may be replaced by new perceptions, new understandings, new "truths."

As you try to answer the questions that follow you will be embarking on a voyage of intellectual exploration. Don't be discouraged if many questions stump you. This book is not a test; it's supposed to be fun. So just sit back and enjoy yourself!

The Sun, The Moon, The Stars, The Universe

Since recorded history began, humans looked up at the stars. What were those lights in the sky? Primitive humans thought the lights might be gods. They watched the skies with wonder. For astronomers, the wonder has never ceased. Where did the sun, the moon, and the stars come from? How big is the universe? Questions like these continue to fascinate and bewilder modern scientific star watchers. The questions that follow will test your knowledge of what humankind has learned about the universe.

1. How old is the universe?
2. How many years old is our solar system?

❸ Astronomers believe that our sun will someday change into a kind of star called a red giant. At that time it will become much larger, expanding outwards to engulf the inner planets (and perhaps the Earth itself). When do astronomers believe this catastrophic event will occur?

❹ *Oh, Be A Fine Girl; Kiss Me Right Now, Sweetheart.* This sentence is a clue to help astronomers remember the different types of stars, in order of decreasing temperature. Thus, from hottest to coolest the star types are O, B A, F, G, K, M, R, N, and S. Our sun is one of these. Which one?

❺ The sun, we now know, is a giant nuclear furnace. What element comprises the bulk of the sun? What element is this "fuel" converted into in the sun's nuclear fires?

❻ Who were the two scientists who detailed the nuclear processes by which the sun converts mass into energy?

❼ You know that all life on Earth depends on energy from the sun. The Earth receives a tremendous amount of solar power. It is estimated that the energy of sunlight reaching the Earth is the equivalent of 170 trillion kilowatts of electricity. That's a lot of juice! What fraction of the sun's total energy output does it represent?

❽ The closest a spacecraft has ever come to the sun is 28 million miles. This close approach was achieved by the sunprobe *Helios II* in 1976. Science fiction writers have dreamed of sending spacecraft into the sun itself, but present-day technology cannot create a ship capable of withstanding the sun's fierce temperatures. So nobody really knows what the inside of the sun is like. Yet scientists believe they do know how hot it is at the core of the sun. What temperatures do scientists believe exist at the sun's core?

9 Like other members of the solar system, the sun also rotates on its axis. How long does it take for the sun to complete one rotation?

10 Since the Earth's orbit around the sun is not a circle, but an ellipse, Earth is not always the same distance from the sun. On the average, however, the distance from the Earth to the sun is about 93,000,000 miles. Astronomers have a name for this distance. What is that name?

11 Light moves at approximately 186,000 miles per second. How long does it take sunlight to cover the 93,000,000 miles between the sun and the Earth?

12 How many times brighter is the light of the sun than the light of the full moon: (a) 1000 times, (b) 100,000 times, or (c) 600,000 times?

13 The lighted area of a full moon is twice as large as a first-quarter or third-quarter moon. Yet the full moon is nine times as bright as a quarter moon. Can you figure out why?

14 We generally think that the moon revolves around the Earth. That's not exactly correct. It's really more accurate to think of the moon and Earth as a "double planet" with each orbiting around a common gravitational center. Where is this gravitational center located?

15 There are presently three scientific theories that attempt to account for the origin of the moon. Theory 1: The moon and Earth both formed from the same cloud of dust and gas and have always occupied more or less their present positions in relation to one another. Theory 2: The moon formed somewhere else in space, and was later "captured" by Earth's gravity. Theory 3: The moon was once part of the Earth and was separated from it at a later time. Which theory is correct?

16 You might be alarmed if you heard that scientists had learned that the moon was falling toward Earth. You shouldn't be. Gravity is always pulling the Earth and moon toward each other, so the moon is falling. Why doesn't it hit the Earth?

17 For a long time scientists believed that the moon had no atmosphere. They thought that the gravitational field of the moon was too weak to hold any gases. But in 1988, researchers were surprised when they detected minute amounts of sodium and potassium gases around the moon. Where does the moon's tenuous atmosphere come from?

18 When we use the phrase "once in a blue moon" we mean "not very often." One kind of "blue moon" is the second of two full moons appearing in the same calendar month—an event that doesn't happen at all in some years. There's another kind of blue moon that occurs when the moon appears to shine with bluish light. What causes this unusual luminosity?

19 Scientists believe that the solar system was formed from a cloud of dust and gas in space, as gravity pulled the atoms in the cloud closer together. What percentage of the mass of the original cloud formed the sun?

20 We usually say that there are nine planets in our solar system. That statement should really say. . . nine *known* planets. One scientist, S. Alan Stern of the University of Colorado, argues that there may be as many as 1,000 as-yet-undetected planets orbiting in the distant reaches of the solar system. If these planets exist, why haven't we located them?

21 Can you name the nine known planets in order of their distance from the sun, starting with the planet closest to the sun? Watch out! There's a bit of a trick to this question.

22 Except for the sun, Jupiter is the most massive object in the solar system. Which do you think has more mass, Jupiter or all the other known members of the solar system combined (excluding the sun)?

23 We are probably all familiar with pictures showing the rings of Saturn. These rings consist of countless billions of small particles that orbit the planet. We now know that two other planets also have ring systems, though not nearly so spectacular as Saturn's. Which other two planets have ring systems?

24 All of the planets in our solar system, except one, spin in the same direction. Which planet spins in the opposite direction?

25 Which planet was thought for many years to be a star?

26 A planet's day is, of course, the time it takes the planet to rotate once on its axis. You might expect that it would be one of the smaller planets that rotates the fastest, but you'd be wrong. The planet with the shortest period of rotation is the largest—Jupiter. How long is a Jovian day?

27 Sometime between November 1990 and July 1991, a geological event occurred that altered the surface of Venus. We've got the evidence, in the form of radar images made by *Magel-*

lan—an exploration module designed to map the Venusian surface. What was the geological event documented by Magellan?

28 At what speed is Earth moving through space in its annual trip around the sun?

29 On what planet in our solar system is a day longer than a year?

30 At its closest approach to Earth, Mars is 34.6 million miles away in a straight line. Yet the Viking landers traveled much farther, because they had to move in a long elliptical orbit to match speeds with Mars. How much farther did the Viking landers travel than the 34.6-million-mile straight line distance?

31 The largest mountain found on any planet is named Olympus Mons (Mt. Olympus) after the legendary home of the Greek gods. This giant, extinct volcanic mountain dwarfs any Terran counterpart. It is 300 miles across at its base; its crater is a whopping 40 miles in diameter. On what planet is this mountain found?

32 How many stars are in our galaxy?

33 Although no scientific instruments have ever been sent far enough to photograph an image of our galaxy from the outside, cosmologists believe that ours is a spiral galaxy, somewhat

like a giant pinwheel with clouds of stars for arms. Our galaxy rotates around its center (which may very well be a black hole). In Earth years, how long does it take the galaxy to complete one rotation: (a) thousands of years, (b) hundreds of thousands of years, or (c) millions of years?

34 When Arno Pensiaz and Robert Wilson, two radio astronomers at Bell Laboratories in New Jersey, picked up some strange radiation on their instruments, they thought the "noise" might be caused by pigeons roosting on their antenna. Later they found that what they were detecting was far more important—important enough to win them the Nobel Prize in physics in 1978. What had their antenna really tuned in on?

35 Some astronomers live more exciting lives than others. The famous Danish astronomer of the sixteenth century, Tycho Brahe, was known as "the man with the golden nose." Why?

36 The great Italian astronomer, Galileo Galilei, was forced to publicly deny that the Earth revolved around the sun—even though his observations proved that it did. But Galileo's "heresy" was not a new belief. The truth had been suggested by a Greek philosopher around 300 B.C. Do you know the name of the Greek philosopher who was the first man known to reason that the Earth revolved around the sun, rather than vice versa?

37 On the night of January 7 in the year 1610, Galileo peered through a telescope at the planet Jupiter. He saw what appeared to be three small stars near Jupiter. The next night, he saw the stars again, but they had moved. Galileo realized they weren't really stars. What were they? Two nights later, only two of the three objects were visible. In a flash of insight, Galileo reasoned that he must be observing satellites in orbit around Jupiter. Fascinated, he watched this spectacle night after night. On the night of January 13, he found yet another moon, for a total of four. Can you name the moons of Jupiter that Galileo discovered? Why were these names given?

38 Galileo was the first astronomer to use a telescope to study the heavens. Did Galileo invent the telescope?

39 Galileo may have suffered severe medical consequences from his astronomical studies. What were they?

40 At a time when women scientists were rare, an American astronomer became the first woman to discover a new comet. She made her discovery in 1847. Later she became the first female member of the American Academy of Arts and Sciences and a professor of astronomy at Vassar College. Who was she?

41 What chemical compound forms the bulk of most comets?

42 Many of the most fascinating objects in the cosmos were catalogued by a French astronomer, Charles Messier, who lived from 1730 to 1817. During the last years of his life, Messier concentrated on the search for comets. To his exasperation, Messier kept finding hazy, blurred objects in the night sky that looked like comets, but were not. (Messier knew they weren't comets because their positions didn't change.) To guide other comet searchers, Messier made a list of these blurry objects. Later, astronomers learned what Messier's noncomets really were. What did they discover about Messier's astronomical rejects?

43 Astronomers believe that ours is a spiral galaxy with arms like a giant pinwheel of stars. That's just one of the two regular shapes galaxies can take. What's the other?

44 The dimensions of our universe are simply staggering to the human imagination. Here's a three-part question to test your understanding of astronomical scale. First, can you name the galaxy that is closest to our own? Second, in light-years, how far away is it? And third, what is that distance stated in miles?

45 Let's get back a little closer to home. Which star is the closest to our own? Measured in light-years, how far away is it?

46 In the year 1929, an American astronomer, Edwin Hubble, figured out an explanation for a fact that had puzzled astronomers. The light coming from distant galaxies showed a red shift—that is, the frequencies of light radiation from the galaxies was shifted towards the red end of the spectrum. Hubble worked out a scientific law by which it was possible to calculate the distance between ourselves and distant galaxies by measuring the red shift. What is this law called?

47 World War II brought an unlooked-for astronomical opportunity to researchers using the 100-inch telescope at Mt. Wilson in southern California. Why was wartime a good time for their studies?

48 As more sophisticated devices are built to scan the secrets of the cosmos, the distance that human scientists can peer into the universe gets greater. Today, radio telescopes are scanning strange objects in the heavens called quasars, some of which are more than 10 billion light-years away. But even the naked eye can see astronomical objects that are unbelievably distant from our planet. What's the most distant astronomical object you can see with just your eyes, and how far away is it?

Answers

① *A scientific estimate (some say "guesstimate") of the age of our universe is about 15 billion years old.*

② *Compared to the universe, our solar system is a youngster—only about 4.5 billion years old.*

③ *Don't worry about being crisped by the expanding sun. Astronomers believe another 5 billion years will pass before it happens.*

④ *Our sun is a G-class star. It has a temperature right in the middle of stellar ranges.*

⑤ *The sun is primarily composed of hydrogen. Nuclear fusion converts this hydrogen into helium—a process during which energy is released.*

⑥ *They were Hans Bethe and Carl von Weizsacker, who worked out their ideas in the 1930s.*

⑦ *Somewhat less than one two-billionth of the energy radiated into space by our sun reaches the Earth.*

⑧ *The best current estimate of temperatures at the sun's core is on the order of 27 million degrees Fahrenheit.*

⑨ *The answer depends on which part of the sun you're talking about. The surface of the sun is not solid—it's gaseous. Near its poles, the sun rotates once every twenty-seven days. Near its equator, the sun rotates once every thirty days.*

⑩ *Astronomers call the distance from the Earth to the sun one Astronomical Unit (often abbreviated AU).*

⑪ *It takes about eight and one-half minutes.*

⑫ *The sun shines about 600,000 times brighter than the full moon, so (c) is the correct answer.*

⑬ *The moon's surface is roughened by mountains, craters, and other geological features. From our point of view, the sun shines "sideways" onto a quarter moon, casting long shadows that dim the moon's brightness.*

⑭ *Because the Earth is so much more massive than the moon, this center of gravity is actually inside the Earth—about 1,000 miles below the Earth's surface.*

⑮ *If you can't choose the correct theory, don't be disappointed. Scientists aren't sure either. All three theories have their advocates, and sure scientific knowledge of the origin of our moon must await further discoveries.*

⑯ *Since the moon is "in orbit" it moves "sideways" just as rapidly as it "falls," it thus remains the same distance from Earth.*

⑰ *It is believed that incoming meteorites vaporize material on the moon's surface, creating a very thin atmosphere (which is constantly leaking away and being replenished).*

⑱ *The moon (and even the sun) can appear blue if certain particles in the Earth's atmosphere (such as those created by forest fires and volcanic eruptions) scatter light at the red end of the spectrum while letting blue light shine through.*

⑲ *Almost all of it. Ninety-nine and eight-tenths percent of the mass of the solar system is locked within the fiery furnaces of the sun. This leaves only two-tenths of one percent to build the planets, moons, comets, and other members of the solar system.*

⑳ *These as-yet-undetected planets are believed to be very distant, very dim "ice dwarfs" that reflect very little light and so cannot be seen through telescopes.*

㉑ *The right answer to this question depends upon when you answer it. Until the year 1999, the correct answer is Mercury, Venus, Earth, Mars, Jupiter, Uranus, Pluto, and Neptune. Usually Pluto is the most distant planet, but its orbit is eccentric, and sometimes it is closer to the sun than Neptune.*

(22) *Jupiter's mass is more than two times as great as that of all other known planets, moons, asteroids, and comets combined in the solar system.*

(23) *Uranus and Jupiter also have ring systems.*

(24) *Venus spins in the opposite direction. Scientists have no good explanation for Venus's strange behavior.*

(25) *Neptune was first observed in 1795 and was thought to be a star. It was identified as a planet in 1846.*

(26) *Jupiter rotates on its axis every nine hours and fifty-five minutes.*

(27) *It was a giant landslide. Images show that a section of a big Venusian cliff fell, forming a patch of rubble beneath. This discovery is a scientific milestone, providing the first evidence of geological change recently occurring on a planet other than Earth.*

(28) *The Earth speeds up and slows down a little during the course of a year; its average orbital speed is close to 67,000 miles per hour.*

(29) *Venus orbits the sun once every 224.7 Earth days. This period of time is a Venusian year. Venus rotates on its axis once every 243.09*

Earth days. This period of time is a Venusian day. Thus, a day on Venus is longer than a Venusian year.

(30) *The Viking landers traversed 440 million miles of space to reach Mars.*

(31) *This volcanic mountain is found on Mars.*

(32) *Nobody knows. Huge portions of the galaxy are hidden by interstellar dust and gas. Estimates, however, place the number of stars in our galaxy at about 100 billion.*

(33) *The correct answer is (c). It takes approximately 230,000,000 years for our galaxy to complete one rotation.*

(34) *The "noise" they were detecting was determined to be radiation left over from the Big Bang, which astronomers believe was the explosion that occurred at the birth of our universe.*

(35) *As a young man, Tycho Brahe fought a duel to maintain the fact that he was a better mathematician than his opponent. Evidently, however, his antagonist was a better swordsman. The astronomer lost much of his nose in the affray. For cosmetic reasons, he had an artificial one constructed—partly of gold.*

(36) *The philosopher was Aristarchus, who lived on the island of Samos in the Aegean Sea.*

(37) *The moons Galileo discovered are Io, Europa, Ganymede, and Callisto. The names were taken from Greek and Roman mythology and are the names of four of the god Jupiter's many lovers.*

(38) *No. Strangely, the arrangement of lenses that makes a telescope work was discovered by two children playing with lenses. In 1608, the children of a Dutch lens grinder named Hans Lipperhey found that by looking through two lenses at once—one a concave lens and the other a convex lens—distant objects were magnified. Excited by his children's discovery, Lipperhey encased two such lenses in a tube and so invented the telescope.*

(39) *Not long before he died in 1642, Galileo went blind. It is believed that this damage may have resulted from his many years of looking through telescopes—particularly from his observations of the sun.*

(40) *This pioneering woman astronomer was Maria Mitchell. Because at that time it was unusual for women to receive formal scientific education, we shouldn't be surprised to learn that Mitchell was taught astronomy privately by her father.*

(41) *Comets consist primarily of water (frozen into ice, of course).*

(42) *The telescopes of Messier's time were not powerful enough for Messier to see*

details of the strange blurry objects. Later astronomers, with more powerful telescopes, discovered that many of Messier's noncomets were galaxies—huge concentrations of stars.

(43) *The other shape galaxies can have qualifies them as elliptical galaxies.*

(44) *The galaxy closest to our own is the Great Spiral Galaxy in Andromeda (also known to astronomers as M31). It is 2.2 million light-years distant from our own. Two and two-tenths million light-years translates into a distance of about 13,000,000,000,000,000 miles!*

(45) *Give yourself half credit if you answered that Alpha Centauri is our closest stellar neighbor. Actually Alpha Centauri is a system of three stars, which averages 4.3 light-years in distance from our sun. Proxima Centauri, which distantly orbits the other two stars of the system, is sometimes the closest and sometimes the farthest of the three. At its closest, Proxima Centauri is 4.2 light-years away across the vastness of space.*

(46) *Logically enough, it is called Hubble's Law. It has enabled astronomers to calculate the dimensions of our universe.*

(47) *Because of the blackout in nearby Los Angeles. Worried about a possible Japanese attack, U.S. authorities ordered lights in the sprawling city to be dimmed and cloaked. The result? A*

night sky unpolluted by stray light from the metropolis, which allowed astronomers to see more clearly.

(48) *The most distant object you see without magnifying lenses is the spiral galaxy M31 in the constellation Andromeda. It is 2.2 million light-years away.*

Mysteries of Matter and Energy

Scientists study what can be seen, felt, heard, smelled, tasted, counted, and measured. Knowing this, would you say that the "stuff" scientists observe and make theories about is basically matter? Or is it basically energy?

There is, of course, no real answer to those questions, because one of the fundamental understandings of modern science is that matter and energy are really the same, as expressed in Einstein's famous equation: $E = mc^2$.

So, in the deepest sense, the two scientific disciplines we call physics and chemistry are really one, although they give us two different windows through which we peer in an attempt to comprehend the universe around us.

So summon your mental energy to the matter at hand as you try to answer the following questions about both the scientists who have probed the mysteries of matter and energy and the fascinating discoveries they have made.

1 It doesn't take a fancy laboratory situation to pose a question about science. Let's consider an everyday event such as a pan of water coming to a boil. When the water gets hot enough, you can see bubbles forming on the bottom of the pan and floating to the surface. What are the bubbles made of?

2 Only one female scientist has ever received two Nobel Prizes. She died as a result of her experiments. Can you name her?

3 Physicists look for answers to such questions as the following: Why is a clear, daytime sky the color we call blue? Do you know why?

4 Physicists strive for ever more accurate means of measurement. Take time, for instance. In 1967, a convocation of world scientists redefined how much time is contained in exactly one second. How did they define a second?

5 Have you ever struggled in frustration as you failed to make catsup pour from a bottle? There's a scientific name for the condition of the catsup that makes it so hard to pour. What is it?

6 One way physics has of honoring its greatest practitioners is to name units of measurement after scientists whose research has led to a deeper understanding of processes going on in the world around us. On the left, in order of the years of their births, you will find twelve scientists who have been commemorated by giving their names to basic units of measurement used in physics. On the right, all scrambled up, is a list of what these units measure. How many of the scientists can you match to the units named after them?

Scientists	Units
1. Newton	**(a)** unit of electro motive force
2. Coulomb	**(b)** unit of electrical charge
3. Watt	**(c)** unit of frequency
4. Volta	**(d)** unit of electrical resistance
5. Ampere	**(e)** unit of electrical inductance
6. Ohm	**(f)** unit of force
7. Faraday	**(g)** unit of energy
8. Henry	**(h)** unit of flux density

9.	Joule	**(i)**	unit of power
10.	Kelvin	**(j)**	unit of capacitance
11.	Tesla	**(k)**	unit of absolute temperature
12.	Hertz	**(l)**	unit of electrical current

7 Let's burn something—say a chunk of wood. While burning, the wood gives off heat and light. It also produces a solid residue of ash, smoke, and soot, as well as releasing gases. If we gathered up all that ash, smoke, soot, and gas, would it weigh more or less than the chunk of wood we burned?

8 There are 96 elements occurring naturally on Earth. Fifteen of these elements are radioactive. Which was the last nonradioactive element to be discovered? When was it discovered?

9 Eight elements make up 98 percent of the Earth's crust. How many of these eight can you name? Which element is the most abundant of all? Which element is the most abundant of the metals?

10 Let's consider Einstein's formula: $E = mc^2$. It implies that matter can be converted into energy, and it tells us just how much energy a given amount of matter can be converted into. Now let's consider one gram (slightly more than 1/28th of an avoirdupois ounce) of matter. Suppose we could convert it into energy and harness that energy to power an ordinary 100-watt incandescent light bulb. How long would the energy "contained" in a gram of matter power that light bulb?

11 Electrical current in a wire is, of course, the flow of electrons from one place to another. Think about an ordinary electrical storage battery. It has a positive pole and a negative pole. If you place a metal conductor between the two poles, electrons flow from one pole to the other. Do they flow from the positive pole to the negative pole or the other way around?

12 A car honking its horn races toward you on a city street. Still honking, it passes you and travels away. You know that the note emitted by the horn doesn't change in pitch, but you hear it as being higher in pitch when the car is moving toward you and lower in pitch as it moves away. Physicists have an explanation for this; it's named after the Austrian scientist who first figured it out mathematically. Can you name him?

13 Mostly, you're just empty space. That's because the atoms you're made of are mostly empty space, or so physicists have determined. Most of the matter in an atom is confined to the nucleus, well over 99 percent of it, in fact. In a "typical" atom, how much of the space in the atom is taken up by the nucleus?

14 You've probably never actually felt sunlight pushing on you, but light does exert pressure. The figure for the pressure exerted by noonday sunlight falling on a square mile of Earth's surface has been calculated. How hard does sunlight push on the Earth?

15 This Greek philosopher said that all matter was made up of tiny indivisible particles that clumped together because each of the particles was surrounded by tiny hooks. He is considered to have been the originator of the "atomic" theory. Can you name him?

16 The British physicist Stephen Hawking is confined to a wheelchair and his speech is intelligible to only his family, a few close friends, and coworkers. His physical problems are due to a rare disease, but there's nothing at all wrong with his mind. He is considered one of the greatest scientists of the twentieth century. What is Hawking's area of greatest expertise?

17 Championship high-divers leap from a thirteen-meter board (more than forty feet above the water). Considering the force of gravity, how fast are these aerial and aquatic daredevils going when they hit the water?

18 Why does a baseball pitcher have to aim high?

19 Even the greatest scientists make mistakes. For instance, Leonardo da Vinci tried to understand the behavior of a body falling toward

the Earth. He reasoned that the speed of the falling body increases in proportion to the distance it has fallen. He was wrong. It would be another Italian, Galileo, who proved that the speed of a falling body increases in proportion to the time it has been falling. The English physicist Isaac Newton would later explain the velocity of a falling body mathematically, in what is called Newton's Gravitational Constant. If you know the speed of fall of a falling body at a specific instant, you can rapidly figure the speed at which the body will be falling if it continues to fall freely for one more second. That is, if you know how to use Newton's Gravitational Constant. Do you?

20 Did Isaac Newton really come to a sudden understanding of gravity when an apple fell on his head?

21 Many people these days are concerned about their weight. Did you know that you can achieve an instant weight loss by moving to a certain place on Earth? Where would you weigh less, at the North Pole or the Equator?

22 A Russian chemist classified the naturally occurring elements into groups. His classification of the elements is called "The Periodic Table" and is still used today—sometimes much to the dismay of students who have to learn it. What was the name of the Russian chemist who classified the elements? For extra credit, can you name any one of the three elements (all unknown at the time) whose existence he predicted and which were later discovered?

23. Molecules are made of atoms joined together. Atoms are composed of particles with such names as protons, neutrons, electrons, and so on. It is now believed that even protons and neutrons are made up of smaller packets of energy/matter stuff. What do physicists call these packets?

24 Cosmology is a branch of physics that deals with the origin and development of our universe. Modern cosmologists believe that the actions of only four different basic forces can explain what's going on in the universe. What are the four basic forces in our universe?

25 On December 2, 1942, an enigmatic phone conversation took place between Professor Arthur Compton in Chicago and James Conant at Harvard University. "The Italian navigator has reached the New World," Compton said. "And how did he find the natives?" Conant asked. "Very friendly," Compton replied. Why was this conversation one of the most important communications in the history of the world?

26 What was the duration of the longest nuclear chain reaction in Earth's history?

27 During the Manhattan Project in World War II, scientists needed something called "heavy water" for their experiments. Heavy water is still used in nuclear reactors. Is it really heavier than ordinary water?

28 In working out their ideas, scientists often rely on conversations with others to stimulate their thinking. Thus, it is said, Albert Einstein discussed his thoughts with his wife, who followed his research very closely. The great nineteenth century physicist, James Clerk Maxwell, had difficulty finding someone to talk with. He was too far ahead of his time. He predicted, for instance, the existence of electromagnetic waves two decades before they were actually detected. So Maxwell talked his scientific problems over with an unusual companion. Who was Maxwell's conversational coworker?

29 In 1887, Heinrich Hertz performed an experiment that revolutionized scientific thinking. What was so earthshaking about his experiment?

30 Electromagnetic waves traveling through space were shocking enough to some scientists in the first years after their discovery. Scientists were even more flabbergasted when Wilhelm Roentgen made an accidental discovery in 1895. What did Roentgen discover?

31 The French scientist Henri Becquerel made an amazing discovery in 1896. It ushered in a new age, and we live in a technological world where developments of his discovery are extremely important. Yet most people don't know what Becquerel discovered. Do you?

32 Who was the first chemist? Nobody knows, of course. Consider that the person who first made a clay pot was a chemist working with clay, water, and fire. Chemistry, after all, is the science of material substances and their changes. Baking a loaf of bread is chemistry. So is the tanning of leather or the dyeing of cloth. Since chemistry is so basic to human affairs it's not surprising that the word *chemistry* refers to one of the world's oldest civilizations. Do you know which ancient culture is referred to by the roots of the word *chemistry*?

33 In the 1860s, French winemakers hired chemist Louis Pasteur to find out why their wine was turning into vinegar. While researching this problem, Pasteur developed a theory that is basic to the practice of modern medicine. What was this theory?

34 All foods spoil eventually—right? Wrong. One food doesn't spoil. Do you know what it is?

35 Citric acid is found in many foods and in citrus fruits in particular. The acid is used as a flavoring and preservative agent in many manufactured foods. What crop supplies almost all the citric acid used in world commerce?

36 The great English scientist, Joseph Priestley, invented new methods of collecting samples of chemically pure gases. He called his process pneumatic chemistry and with it he isolated a gas he called dephlogisticated air. By what name do we know dephlogisticated air?

37 Joseph Priestley made many fundamental chemical discoveries. One of them is used today by people all around the world. You have probably consumed examples of Priestley's chemical wizardry. What kind of beverage did one of Priestley's discoveries make possible?

38 What chemical property of the element carbon makes it the basis of life as we know it?

39 You know what sodium bicarbonate is. It's baking soda, of course. Sodium chloride? Ordinary table salt. Hydrogen dioxide? That's water. What great French chemist began the modern scientific practice of naming chemical compounds by the elements that compose them?

40 Chemistry was simple for students of the Greek philosopher Aristotle. They had to learn the names of only four elements—earth, air, fire, and water. Aristotle believed and taught that all things were composed of varying proportions of these four primordial substances. Today we know there are more than 100 elements. Which Irish scientist pro-

posed the basic theory of our modern understanding of elements?

41 One of the most fundamental kinds of chemical tests is performed with litmus paper. When strips of litmus paper are dipped into a solution, their changing colors indicate whether the solution is an acid or a base. Litmus paper can do this because of a naturally occurring dye found in what kind of plant?

42 Most of the plastics we use are based on carbon. For a long time, doctors tried to implant various carbon-plastic devices in their patients' bodies. This led to a chemical problem. Too often, the body's immune system would reject the implant. Doctors had to turn to plastics based on an element other than carbon. Which element?

43 How many different chemical compounds have been discovered by chemical science?

44 About 60,000 chemicals are used by science and industry in the United States. These days, there is much concern about the health effects of chemicals. What percentage of chemicals used in industry have been tested for toxic effects?

45 For thousands of years scientists knew of two forms of pure carbon. One is graphite (familiar in the "lead" of pencils), and the other is the glittering substance we know as diamond. A new form of pure carbon was discovered in the 1980s. It is a stable aggregation of exactly sixty carbon atoms. Chemists named the new material *buckminsterfullerene.* Why did they pick such an unusual name?

46 Mix a certain white metal that burns in water with a yellowish-green poisonous gas and you get an edible mineral. What is it?

47 During the Napoleonic wars, officials of the French government roamed the countryside. You might have thought they were crazy. These officials went around actually tasting piles of manure and human refuse. What were they up to?

48 We don't usually think of it as chemistry, but the refining, alloying, and working of metals is a chemical process, and so it is an old branch of the chemical industry. Which metal was the first to be utilized by human craftspeople?

49 Before humankind mastered the working of iron, the hardest metals available were alloys of copper. Two of these alloys have been used for thousands of years. They are bronze and brass. What other metal must be mixed with copper to make these two alloys?

50 Let's see how much you know about the chemistry of the air you breathe. There's a certain substance found in every lungful that comes in two different packages. Packaged one way, this substance is essential to life. Packaged the other way, this substance is a pollutant and poison, and yet it is also essential to life on Earth. Name this substance.

51 You're hiking through the Adirondacks when you come upon a group of scientists dumping sacks of lime into a remote mountain lake. What are they doing? Are they feeding the fish? Are they trying to stimulate the growth of aquatic plants? Are they attempting to correct the effects of acid rain?

52 You've heard of the "greenhouse effect" caused by certain gases in the atmosphere. Scientists worry that buildup of greenhouse gases may raise worldwide temperatures rapidly, causing environmental havoc. Which greenhouse gas worries scientists the most?

Answers

① *Actually, the first bubbles to appear before the water reaches the boiling point are the tiny bubbles of air that form along the sides of the container. The larger steam bubbles come later.*

② *She was the Polish-born scientist Marie Sklodowska, better known to history as Madame Curie. Marie and her husband, Pierre, isolated and named two new radioactive elements—polonium and curium. For this, they were jointly awarded the Nobel Prize in Physics in 1903. In 1911, Marie Curie was given the Nobel Prize in Chemistry (her husband had died some years earlier). In 1934, Madame Curie died of leukemia caused by long exposure to radioactivity.*

③ *Because molecules in the atmosphere bend and scatter blue light. Thus blue light, which is not heading toward your eyes as it comes from the sun, is bent toward you while other colors of light proceed to other destinations.*

④ *The scientists decided that a second could best be measured by counting the vibrations of an atom of cesium. They decreed that one second is the period of time it takes a cesium atom to perform 9,162,631,770 oscillations. So you don't have to count "one-thousand one, one-thousand two. . ." any more. Just watch an atom of cesium very closely.*

⑤ *You can astound your friends by informing them that catsup is "thixotropic," which means it's neither exactly a liquid nor a solid. Knowing this, however, does not help get catsup onto your hot dog any faster.*

⑥ *The answers are 1(f), 2(b), 3(i), 4(a), 5(l), 6(d), 7(j), 8(e), 9(g), 10(k), 11(h), 12(c).*

⑦ *It might surprise you, but the solid materials and gas produced by burning weigh more than what is burned. This is because chemicals in the burned material have combined with oxygen in the air.*

(8) *The last nonradioactive element to be discovered was rhenium—a rare, silver-white, very hard metal. It was discovered in 1925.*

(9) *The eight most common elements found in the Earth's crust are oxygen, silicon, aluminum, iron, calcium, sodium, potassium, and magnesium. Oxygen is the most abundant—making up nearly half the Earth's crust. Aluminum, which constitutes one-twelfth of the crust, is by far the most abundant of the metals.*

(10) *Incredibly, the energy in one gram of matter, if used at 100 percent efficiency, would power that light bulb for nearly 30,000 years!*

(11) *No. Electrons do. By convention, current (carried by either kind of charge) flows from positive to negative.*

(12) *He was Christian Doppler, and the phenomenon is called the Doppler Effect.*

(13) *Incredibly, the "typical" atomic nucleus occupies only about one-quadrillionth of the volume of the atom, even though it contains almost all the mass.*

(14) *Calculations show that sunlight pushing on a square mile of Earth's surface does so with a force of about two pounds.*

(15) *He was Democritus, who lived in the fifth century* B.C.

⑯ *Hawking's greatest contribution to our understanding of the universe is undoubtedly his work on the physics of black holes.*

⑰ *Careful measurement and calculation has revealed that people diving off a thirteen-meter board hit the water at about forty-seven kilometers (twenty-nine miles) per hour. No wonder injuries occur if they hit wrong!*

⑱ *Because, as Isaac Newton said, gravity doesn't go away. All the time a baseball is traveling toward home plate, it is falling toward the ground. By the time it smacks into the catcher's mitt, even a fastball dips about two and a half feet from the height at which the pitcher released it.*

⑲ *Newton determined that, in Earth's gravity, a falling body accelerates at a rate of thirty-two feet per second. Thus if you know the speed of the falling body at one point, you can predict its speed one second later by adding thirty-two feet per second to the first figure.*

⑳ *Very unlikely. Newton himself said that the idea came to him when he simultaneously saw an apple hanging on a tree and the moon moving through the sky. He wondered if the apple and the moon both obeyed the same laws of nature.*

㉑ *You would weigh less at the Equator, because gravity is less there. How much less? Well, a 150-pound person would weigh about seven ounces less at the Equator than at the North Pole. Brrr. . . now you'd better put your clothes back on.*

㉒ *His name was Dmitri Mendeleev (also spelled Mendeleyev). The three elements whose existence he predicted are gallium, scandium, and germanium. (If you knew all that, you're a real chemistry whiz!)*

㉓ *These small packets are called quarks. Nobody has ever seen one, but scientists puzzling about how our universe works believe they must exist.*

㉔ *The four basic forces are gravity, electromagnetism, the strong nuclear force, and the weak nuclear force.*

㉕ *Compton was telling Conant in code that a team of researchers led by Italian Enrico Fermi had successfully produced a controlled nuclear reaction. This meant that humankind could now unlock the power of the atom for peace and war.*

㉖ *About 100,000 years. This chain reaction happened about a billion years ago in rich uranium deposits in Africa.*

㉗ *Yes, it's about 10 percent heavier. The hydrogen in heavy water contains an extra neutron in its nucleus. About one hydrogen atom in 5,000 contains this extra neutron.*

(28) *Maxwell talked out his research with Tobi, his dog, who of course understood every word. Maxwell never revealed whether or not Tobi helped him with the complicated mathematics.*

(29) *Confirming Maxwell's theoretical speculations, Hertz's experiment demonstrated that energy can be transmitted through space in the form of electromagnetic waves.*

(30) *Roentgen discovered X rays. The ability of X rays to pass right through solid matter astounded the scientific community.*

(31) *Becquerel discovered that some naturally occurring minerals were radioactive when an ore containing uranium accidentally exposed a photographic plate in a drawer in his laboratory.*

(32) *The civilization is that of ancient Egypt.* Khemia *is an Arab word for the Land of the Nile.*

(33) *It was the germ theory of disease. Pasteur proved that wine was turning into vinegar because of infection by airborne microorganisms. He later proved that many human diseases were caused by the action of germs and bacteria.*

(34) *It is honey, which has natural antibacterial properties. Honey recovered from Egyptian tombs thousands of years old was found to still be edible.*

(35) *The citric acid of commerce is not obtained from any edible crop. It comes from a common black mold,* Aspergillus niger, *which grows naturally on rotting vegetation.*

(36) *Dephlogisticated air was Priestley's name for oxygen.*

(37) *Four years before some of the British colonies in North America declared independence, Priestley described a method of forcing carbon dioxide into water, and thus invented the carbonated beverage industry.*

(38) *Carbon molecules readily join together into long chains and rings, thus making possible the complex substances necessary for life.*

(39) *Antoine Lavoisier is considered the father of modern chemical names. Tragically, this great scientist lost his head in the French Revolution.*

(40) *The Irish scientist who developed the modern theory of chemical elements was Robert Boyle, who lived from 1627-1691.*

(41) *The dye in litmus paper comes from lichen.*

(42) *The plastics used in most surgical implants are based on silicon.*

(43) *There's no exact answer to this question because new chemical compounds are*

constantly being discovered and created. But the latest count identifies about 5,000,000 chemical compounds.

(44) *Only about 30 percent. The health effects (if any) of the other 70 percent are unknown.*

(45) *The name* buckminsterfullerene *was chosen because the shape of the new molecule is similar to that of geodesic domes invented by the American engineer Buckminster Fuller.*

(46) *The edible mineral is table salt, which is composed of sodium and chlorine.*

(47) *They were looking for sources of potassium nitrate—also known as saltpeter—which is an essential ingredient in the manufacture of gunpowder. A British naval blockade had cut shipping to and from France, forcing the French to exploit these unsavory sources.*

(48) *Research into antiquity indicates that gold was the first metal to be worked by humans, as long ago as 5000* B.C.

(49) *Brass and bronze are both mixtures of copper and tin.*

(50) *The substance is oxygen. Most free oxygen in the atmosphere is what chemists call* O_2*. That is, it consists of two atoms of oxygen joined together. This is the oxygen we extract from the air we breathe. A small percentage of oxygen, however, consists*

of single atoms of oxygen. This kind of oxygen is called ozone. When breathed, ozone is a poison, but high up in the Earth's atmosphere the ozone layer protects all life on Earth from damaging ultraviolet radiation.

(51) *They are trying to correct the effects of acid rain. Adding lime to water contaminated by acid rain has been found to reduce the acidity that kills fish.*

(52) *The gas contributing the most to the greenhouse effect is carbon dioxide. Every year, the burning of fossil fuels dumps more than five billion tons of carbon dioxide into the atmosphere!*

Of Wind and Weather

"Red sky at night, sailor's delight; red sky at morning, sailors take warning."

Once upon a time wise words such as these guided human attempts to guess what kind of weather was coming next.

In today's world of technological advances, we have learned a lot more about forces that drive the planetary swirl of the wind, water, sun, cloud, and shadow we call weather. Many thousands of weather stations scattered across the globe contribute data to meteorologists trying to understand and predict the weather. Above Earth's atmosphere, in the weatherless void of space, satellites bearing sophisticated monitoring devices scan the seething air below; their data, too, is pondered by meteorologists. Complex computers are used to analyze patterns of heat and cold, humidity and dryness.

And yet, with all our modern capabilities, the men and women who attempt to predict the weather often can't tell you if it's going to rain on your parade.

So don't be surprised if some of the following questions stump you. The experts don't totally understand weather either.

1 Let's start out with a basic question about Earth's weather. What is the peculiarity of the Earth's orbital position that accounts for the fact that it never snows on Christmas Day in Australia?

2 The Earth is not always the same distance from the sun. During winter in the

Northern Hemisphere, the Earth is actually more than 3 million miles closer to the sun than it is in the summertime. The mathematical word for the kind of orbit the Earth follows around the sun is not a circle. What is the correct word for the shape of Earth's orbit?

❸ Summer in the Northern Hemisphere is the time of year when the tilt of the Earth exposes that part of our planet to long days of sunshine. What is usually the longest day in the Northern Hemisphere? What is another name for the longest day?

❹ In the United States, the Fahrenheit scale of temperature measurements is used. The Fahrenheit scale has been around a long time. In it, at sea level, the freezing point of water is 32 degrees and the boiling point of water is 212 degrees. Those people who live in the United States are used to the Fahrenheit scale of temperatures. When the outside temperature is 100 degrees (Fahrenheit), it's a hot day. Most of the rest of the world uses the Celsius scale of temperature measurements. In it, water freezes at 0 degrees and boils at 100 degrees. So when people discuss the weather in a country that uses the Celsius scale, what's a hot day? Is 38 degrees hot?

❺ The Celsius temperature scale was invented by a Swedish scientist and experimenter, Anders Celsius, in the middle of the nineteenth century. What was Celsius's main field of science?

6 There is a third kind of temperature scale used in meteorology and in other sciences. It defines zero as that temperature at which all molecular motion ceases. This "zero point" is equivalent to a temperature of −459.72 degrees Fahrenheit. That is cold indeed! What British scientist invented this third temperature scale (it is called by his name)? What's another scientific word for that temperature at which all molecular motion ceases?

7 If you look at a globe or a map of the Earth, you can't fail to be impressed by Antarctica, which is always shown in white, for a very good reason—Antarctica is almost completely covered with ice. You might guess that it's cold in Antarctica, and you would be right. As a matter of fact, the coldest temperature ever recorded on Earth was measured by a team of Russian scientists in Antarctica in 1960. How cold was it?

8 The coldest temperature ever recorded on the North American continent was obtained at Skag, Yukon, in February of 1947. Was the temperature (measured in Fahrenheit degrees) warmer or colder than 80 degrees below 0?

9 Let's go from the record cold temperature to the record high temperature. This time I'll give you the highest temperature ever recorded on the North American continent—it was 134 degrees (Fahrenheit). That's a sizzler! Where did this record-breaking air temperature occur?

❿ Our fiftieth state, Hawaii, is in the middle of the Pacific Ocean. Naturally, being surrounded by water has a lot to do with Hawaii's pleasant climate—not often too hot or too cold for comfort. Thinking in Fahrenheit degrees, do you think that the temperature ever fell below 0 degrees in Hawaii? Did it ever rise above 100 degrees?

⓫ A system of mountain chains runs north to south down the western part of the North American continent. In the United States, the western slopes of these mountains are green, forested, and well-watered. The eastern slopes of the same mountains are often dry and stretch away to dry deserts beneath. Why does it rain on one side of the mountains more frequently than it does on the other side?

⓬ There is a mountain rated by meteorologists as having "the world's worst weather." The highest wind (not associated with a tornado or a hurricane) ever recorded on Earth blew on this mountain. Instruments in a weather station atop the mountain recorded a wind speed of 234 miles per hour! This storm-battered peak is not in one of the big mountain chains such as the Himalayas, the Andes, or the Rockies. What is the name of the mountain and where is it located?

⓭ Another mountain, also in the United States, holds the record for "the world's wettest spot." In what state is this rainy mountain located?

14 Here's a question that has an alarming answer. First, let's consider what "clean air" is. You might be surprised to know how many particles of dust there are floating around in the air. Even the cleanest air samples ever analyzed were found to contain more than 500 particles of dust in every cubic centimeter of air! These samples were taken in the middle of the Pacific Ocean. Of course, on land, where wind blows up soil, and in cities, where man-made pollution spews particles into the air, the particle count goes way up. Do you think the dirtiest air ever analyzed contained (a) about ten thousand particles in every cubic centimeter, (b) about one hundred thousand particles per cubic centimeter, or (c) about one million particles per cubic centimeter?

15 Here's a simple two-part question about raindrops. You've surely experienced enough rain to know that raindrops come in lots of different sizes—sometimes big, cold ones that sting when they hit, and sometimes fine, misty ones you can barely feel. Scientists have devoted a lot of study to raindrops and have actually measured their sizes. How little are the littlest raindrops? How big are the biggest?

16 How many atoms are there in an "average" raindrop?

17 Think about raindrops falling out of the sky. On the average, how fast do you think a raindrop is falling when it hits the ground?

18 How much rain do you think could fall in a one-minute period? What is, so far, the record sixty-second downpour that occurred in Maryland in the summer of July 1856? Was the champion one-minute rainfall more or less than an inch of water?

19 While on the subject of big rains, let's consider a time period of one hour. The record amount of rain that ever fell inside of an hour actually took only forty-two minutes to pour from the heavens. This rainfall occurred in Missouri in 1947. How much rain fell in those forty-two minutes?

20 One more record rainfall question. What's the greatest amount of rain ever measured in a single day?

21 The most powerful storms in the world are hurricanes. These storms are born over warm ocean waters as water evaporates from the sea and rises into towering, swirling systems of clouds, winds, and rain that have awesome destructive power. Scientists estimate that nine-tenths of the energy in a hurricane is released to build the clouds that form its familiar pinwheel shape. Only three percent of the energy of a hurricane is converted into a weather aspect of hurricanes that we usually think of first—their winds. But that three-percent wind energy causes havoc when it encounters ships at sea or buildings on land. How fast do the peak winds of a hurricane blow?

22 To get some idea of the total energy in a hurricane, let's consider this question: If all the energy in a hurricane could be converted to electricity, would it be greater or less than the amount of electricity used in the United States?

23 You may know that the word *hurricane* was borrowed from Indian languages of peoples living in the Caribbean and Gulf of Mexico when white men came. We still use the Indian word for tropical storms that are born over the Atlantic Ocean. What are similar storms called in the Pacific Ocean?

24 In the Northern Hemisphere, do hurricanes (like most other storms) circulate clockwise, or counterclockwise?

25 What's the biggest snowstorm you ever saw? You'll have to answer that question for yourself, but let's ask a question about a winter storm that set a record that has stood for more than seventy years. It's the record for the greatest amount of snow ever to fall in a twenty-four-hour period. How much snow fell?

26 Right now, how many thunderstorms do you think are booming their way over the face of the Earth?

27 Which section of the United States has the most thunderstorms per year?

28 You might not be wrestling with the answers to these questions if it weren't for lightning. What does lightning have to do with the presence of life on Earth?

29 Do most lightning strokes move from cloud to cloud, from cloud to ground, or from ground to cloud?

30 Strange tales were told by the earliest settlers of North America's Great Plains states. They spoke of "ball lightning"—mysterious blobs of electrical energy a few inches in diameter that played odd tricks. "Ball lightning" was seen to roll in one door of a house and out another; it was observed following the course of barbed-wire fences. It sometimes startled watchers who saw it floating through the air, even against the wind. For a long time, scientists said "ball lightning" could not possibly exist. What do present-day researchers say about the matter?

31 At what speed does lightning travel through the air?

32 Which is hotter, the air around a lightning bolt or the surface of the sun?

33 Lightning never strikes twice in the same place. True or false?

34 Did you ever see any really big hailstones? Hail as big as five inches across has been reported in many parts of the world. As you can imagine, balls of ice five inches in diameter can do real damage when they fall from the sky. And yes, their impact has actually killed people. The greatest loss of human life due to a single hailstorm occurred in 1888, in northern India. Two hundred and forty-six people died. Some of the victims were simply battered to death. Others died from another cause. Can you guess what it was?

35 No other weather phenomenon can wreak such destruction as the swirling vortex of wind we call a tornado. In terms of human life, the worst tornado on record struck the states of Missouri, Illinois, and Indiana, in March of 1925, killing 689 people. What makes tornadoes so deadly is the speed at which their funnel rotates. What's the fastest that tornado winds can blow?

36 What kind of instruments do weather scientists use to measure the speed of winds in a tornado?

37 Buildings in the storm path of a tornado often appear to have "exploded"—that is, the windows, doors, and even the walls of the buildings blow outward. There's a simple meteorological explanation for this fact. What is it?

38 The first weather scientist in the United States was the Reverend John Campanius, who came to this country to tend his flock of

Swedish immigrants in the mid-1600s. His careful records of weather are the oldest known for the North American continent. Where in the United States did Reverend Campanius make these pioneering weather records?

39 In 1743, a famous American statesman who was also a scientist, made the first recorded study of a hurricane's motion across the Earth. Who was he?

40 During the American Revolution, two men in Virginia carried out a series of scientific weather observations. They made weather observations at the same time at two locations many miles apart. Both of these men would be presidents—one of the United States of America, the other of a notable university. Who were these two future presidents?

41 In 1776, when he was writing the Declaration of Independence, Thomas Jefferson is known to have purchased two different scientific weather observation instruments. The instruments he bought were "state-of-the-art" at the time, though meteorologists today would consider them crude and primitive. Yet they measured two weather statistics that are still taken and kept by weather researchers today. What were the two basic meteorological instruments that Jefferson bought?

42 Who were the first weather observers to make a scientific study of weather in the area that would become the far western United States?

43 In the first part of the nineteenth century, an American scientist was the first man to understand why clouds form in the atmosphere. He stated that clouds are born when air rises and becomes cooler, causing moisture in the air to condense into water droplets. Do you know the name of this famous scientist?

44 Early weather observers were hindered in their efforts to understand the weather because they had no way of taking weather readings high above the ground, where most weather takes place. Starting in 1898 and ending in 1933, the U.S. Weather Bureau used what is usually considered a child's toy to obtain scientific information about weather processes going on in the sky. What was the "toy" that provided almost a half century of valuable scientific data?

45 The modern age of aeronautics vastly expanded scientists' ability to understand weather, since balloons, aircraft, and rockets enabled them to lift their measuring instruments far above the Earth's surface. In search of a better understanding of weather, two daring U.S. military aviators became the first to fly into (and back out of) a hurricane. This bold meteorological flight occurred during wartime. Did this first airplane flight through a hurricane take place during (a) World War I, (b) World War II, or (c) the Vietnam War?

46 If you are curious about weather, you probably watch weather reports on TV, or you may even look in a newspaper for daily weather maps and temperature charts that tell you about weather conditions all over the world. Which newspaper was the first to publish regular weather maps and what was the year?

47 It's not only ordinary people wondering if their picnic will be rained on who rely on weather forecasts. Captains of ships at sea, navigators of aircraft, people who direct construction projects, commanders of military operations, farmers considering the planting or cultivation of crops—all these people and more depend on weather information to make intelligent choices. When were the first regular weather information radio broadcasts made and what was their purpose?

48 If you're watching a weather report on TV and you hear that "the skies are clear," does that mean that there are absolutely no clouds in the sky?

49 One of the statistics routinely given by weather reporters is the humidity. The higher the humidity the more moisture there is in the air, of course. But, scientifically speaking, do you know what "humidity" really means? For instance, can the humidity exceed 100?

50 Did you ever see a big ring like a rainbow around the moon? It's a scientific fact that the angle between the moon, the person observing the ring, and the ring itself is always the same—28 degrees. Why is this so?

51 Now for a final question. You know that meteorites are constantly bombarding Earth's atmosphere. Do they have any effect on the weather we experience?

Answers

1 *It's the twenty-three and one-half degree tilt the Earth has in relation to the plane of its orbit around the sun. This tilt accounts for Earth's seasons, and puts December in the middle of Australia's summer.*

2 *The shape of Earth's orbit is called an ellipse.*

3 *The longest day of the year in the Northern Hemisphere is usually June 21. This day is also called "the summer solstice."*

4 *Yes, if you're talking Celsius, 38 degrees is hot. It's the equivalent of a Fahrenheit temperature of slightly over 100 degrees.*

5 *Anders Celsius was a Swedish astronomer.*

⑥ *The scientist was the famous British mathematician and physicist, Lord William Kelvin. The temperature at which all molecular motion ceases is also called "absolute zero."*

⑦ *The record low temperature measured by the Russians in Antarctica was a frigid 128.94 degrees (Fahrenheit) below 0.*

⑧ *It was colder. The temperature recorded at Skag, Yukon, was measured at 81 degrees (Fahrenheit) below 0.*

⑨ *The temperature of 134 degrees (Fahrenheit), was recorded in Death Valley, California, in 1913. This dry desert valley is actually lower than sea level. The height of atmosphere above it accounts for the high temperatures recorded there. The 134-degree reading just missed being a world record. That record temperature, of just over 135 degrees, was set in 1922 in the deserts of northern Africa.*

⑩ *The temperature has never fallen below 0 degrees in Hawaii. The coldest reading ever taken was 14 degrees above 0. Official thermometers have never gone above 100 degrees in Hawaii. The record Hawaiian heat wave measured exactly that temperature—100 degrees Fahrenheit.*

⑪ *The answer is really quite simple. Winds in the western United States generally blow in from the Pacific Ocean, bearing moisture that forms into clouds as air moves up the mountain slopes. The clouds produce rain; so the western slopes of the mountains are green. By the time the air crosses over the mountains, it is as if all the water has been wrung out of it. Thus, little rain falls east of the western mountains, producing desert conditions.*

⑫ *The worst-weather champion is in the United States. It's Mt. Washington, which is located in the White Mountains of New Hampshire.*

⑬ *In Hawaii. Weather records show that Mt. Waialeale, over a period of thirty-seven years, received slightly more than 471 inches of rain a year. That averages out to well over an inch a day!*

⑭ *The answer is alarming enough, but it's not the biggest of those three figures. The dirtiest city air can contain slightly more than one hundred thousand particles per cubic centimeter of air, so the correct answer is (b).*

⑮ *The smallest raindrops are about one one-hundredth of an inch in diameter. The biggest are about one-quarter of an inch in diameter.*

⑯ *The average raindrop is said to contain 6,000,000,000,000,000,000,000 atoms; approximately two-thirds are hydrogen atoms and one-third are oxygen atoms. If you're wondering how to say the name of that number, it's 6 sextillion.*

⑰ *The average speed of falling raindrops has been determined to be approximately seven miles per hour.*

⑱ *This record one-minute rainfall measured more than an inch—one and twenty-three hundredths of an inch, to be exact.*

⑲ *The forty-two-minute downpour in Missouri was measured at an even twelve inches of rain.*

⑳ *The record one-day rainfall is an incredible 73.62 inches. That's more than six feet of water! It occurred in 1952 on La Reunion, an island in the Indian Ocean.*

㉑ *Hurricane winds have been clocked at speeds of more than 200 miles per hour.*

㉒ *This contest is not even close. The wind energy generated by an average hurricane during its few-day lifespan is equal to many years of electrical usage in the United States.*

㉓ *Hurricane-like storms in the Pacific Ocean are called typhoons, from Chinese words meaning "great wind."*

㉔ *Hurricanes in the Northern Hemisphere circulate in a counterclockwise direction.*

㉕ *On April 14 and 15 of 1921, in twenty-four hours, more than six feet (75.8 inches) of snow fell on Silver Lake, Colorado.*

(26) This kind of question can't be answered exactly, of course. But a good bet would be that, at this instant, there are approximately 2,000 thunderstorms in progress. That adds up to 16,000,000 thunderstorms a year on our planet. That's a lot of lightning!

(27) The southeastern United States has the greatest number of thunderstorms per year. Sections of Florida may have more than ninety thunderstorms in an average year.

(28) Many scientists believe that life on Earth originated when the electrical charge of lightning caused a chemical change in a primeval soup of gases—ammonia, methane, hydrogen, and water vapor—that enveloped the planet. It has been demonstrated in the laboratory that such an event could have created amino acids, which are the fundamental building blocks of proteins. Proteins are, of course, essential to all forms of life that we know about.

(29) All three of these kinds of lightning strokes occur. The most common kind discharges electricity from clouds to the ground.

(30) The highly-charged scientific dispute over whether "ball lightning" exists has been resolved in favor of the proposition that it does exist. Scientists believe that "ball lightning" consists of superheated regions of air (temperatures may range above 10,000 degrees Fahrenheit) that possess high electrical conductivity.

(31) *The speeds attained by lightning strokes are truly amazing. In a typical cloud-to-ground discharge, the first propagation of electrical energy may move at the rate of only 100 miles per second. But when what is called "the return stroke" shoots upward from the surface of the Earth, its speed may exceed 80,000 miles per second!*

(32) *Temperatures associated with lightning bolts are far higher than temperatures on the surface of the sun. Scientists have measured air temperatures around lightning bolts as high as 50,000 degrees Fahrenheit. The surface of the sun is "only" 10,000 degrees Fahrenheit or so.*

(33) *Lightning often "chooses" to strike in the same place again and again. Tall trees and buildings are favorite targets. For many years the Empire State Building in New York City was the tallest building in the world. During one storm, lightning struck the skyscraper no less than twelve times. Modern buildings are designed, of course, so that the electrical energy in lightning is channeled harmlessly into the ground.*

(34) *They froze to death. The hailstorm was so intense that victims were buried beneath "drifts" of hailstones.*

(35) *It is believed that wind speed in tornadoes can exceed 500 miles per hour.*

(36) *Ordinary weather instruments do not withstand the onslaught of tornado winds. Estimates of the strength of tornadoes have been made by analyzing destructive effects such as those wrought by a tornado that swept through St. Louis, Missouri, in 1896. During this storm, winds picked up an ordinary two-by-four and drove it through a sheet of iron five-eighths of an inch thick.*

(37) *The center of a tornado is a vortex of extremely low air pressure. When this low-pressure area passes over a building, air trapped inside the building pushes the walls and windows outward.*

(38) *Campanius made weather measurements near what is now Wilmington, Delaware, in the years 1644 and 1645.*

(39) *None other than Benjamin Franklin, who did far more than fly a kite in a storm and invent bifocals. Franklin investigated the times at which a hurricane reached and passed over Philadelphia and then New York City. He determined that the storm was moving up the northeast coast of the United States in a northeasterly direction.*

(40) *The two men were Thomas Jefferson (later, of course, president of the United States) and James Madison (president of William and Mary College). You will notice that the second of these men had the same name as another Virginian who*

would become president of the United States, although he was not the same man.

(41) *They were a thermometer and a barometer. Temperature and air pressure are basic measurements that help describe Earth's weather.*

(42) *They're usually thought of as explorers, not scientists, but Meriwether Lewis and William Clark, who explored the Louisiana Purchase in the years 1803-06, carried sophisticated surveying and weather observation instruments. Their journals, maps, and drawings contain a great deal of scientific information.*

(43) *This famous American weather scientist was James Espy. In 1842, Espy became the first officially-appointed weather scientist to work for the U.S. government.*

(44) *The method used to gain information about atmospheric conditions high above the Earth was to attach scientific instruments to kites and fly them as high as possible.*

(45) *The correct answer is (b). It was 1943 when Colonel Joseph P. Duckworth and Lieutenant Ralph O'Hair made their bold aerial pass through a hurricane, obtaining weather measurements as they flew.*

(46) *The first newspaper weather maps appeared in the* New York Graphic *in 1879.*

(47) *The first regularly-scheduled weather broadcasts began in 1902. Weather information coded into telegraphic form was broadcast by radio waves to luxury Atlantic liners so that they could avoid storms and head winds.*

(48) *No. "Clear sky" can actually be as much as one-tenth covered with clouds.*

(49) *No. Relative humidity—often called the "humidity index"—is a percentage. It is a measure of the amount of water in air of a given temperature, expressed as a percentage of the maximum amount of water that air of that temperature could contain. So a humidity index of 100 means that the air is completely saturated with water. If more water is added to the air, it will condense on objects on the ground or fall as rain.*

(50) *The correct answer has to do with the prismatic qualities of tiny ice crystals high in the Earth's atmosphere. If you see a ring of colors around the moon, you are seeing moonlight broken into colors by countless numbers of ice particles in the sky. The physical laws of ice crystal formation and the nature of light determine that the ring must be exactly 28 degrees from the moon.*

(51) *This would be a good question to ask if you wanted to start an argument at a convention of meteorologists. Many weather scientists*

believe that their studies show effects of meteor showers such as an annual pattern of increased rainfall associated with meteors. Other analysts say there's nothing to it.

Our Watery World

The continents we live on are really islands, surrounded by the great world oceans that cover almost three-quarters of our watery world.

The scientific study of the oceans and ocean creatures is called oceanography or oceanology. In surface ships, submarines, and laboratories around the world, scientists seek an understanding of what goes on in the 330 million cubic miles of water contained in the oceans.

Test your knowledge of what they have discovered.

1 How many water molecules are in a single drop of seawater?

❷ Gravitational attraction by the sun and moon cause a phenomenon we call the tides—the twice daily ebb and flow of water in the world's oceans. In small, landlocked seas such as the Mediterranean, the tides are only a few inches in height. But, where they are funneled by landforms, the tides run many feet high. Where do the highest tides in the world occur?

❸ The ocean tides are said to "ebb" and "flow." Is ebb tide the same as high tide or low tide?

❹ A tsunami is (a) an Asiatic fish, (b) a high wind of the South Seas, or (c) a huge ocean wave?

❺ Does sound travel faster or slower in water than it does in the air?

❻ Pouring north past the east coast of Florida and the Carolinas flows one of the mightiest "rivers" in the world, a river in the ocean—the Gulf Stream. It affects climates in England, Ireland, France, and Scandinavia, keeping these lands warmer than they would otherwise be. The Gulf Stream has changed history more than once. The Pilgrims who landed in what is now Massachusetts were supposedly trying to reach Virginia. The Gulf Stream, about which sailors were ignorant at the time, may have pushed them to the north. As more and more ships sailed back and forth between Europe and North America, navigators learned about the Gulf Stream. Sailing with or against its current could cut or add many days to a transatlantic journey. Naturally, many merchants and ships' captains wanted to know the exact course of the Gulf Stream. Can you name the famous American politician and scientist who produced the first map that accurately plotted the course of this ocean river?

7 Ponce de Leon, the Spanish explorer who looked for the Fountain of Youth, had a strange experience off the east coast of Florida. Although there was a strong wind in his ships' sails, they could not move ahead in the water. What happened?

8 Which country of the world has the longest coastline?

9 In a channel off the Greek island of Euboea in the Aegean Sea, the current reverses direction fourteen times a day. It is said that this strange current caused the death of Aristotle. How?

10 In the nineteenth century, the captains of many U.S. ships put messages in bottles and threw them overboard. In doing that they were making an important contribution to ocean science and navigation. Can you guess what was written in the messages?

11 Imagine a ship above the Marianas Trench in the South Pacific. The ship is riding on 36,000 feet of water, which means that the deepest part of the trench is almost seven miles down. If someone dropped a stone overboard, how long would it take to reach the floor of the trench?

12 Scuba diving is a popular sport and is also a way for scientists to explore below the surface of the sea. Scuba divers usually enter the water backwards. Why?

13 Most kinds of limestone are made from deposits that accumulate on the bottom of the sea. What do these deposits consist of?

14 Most kinds of sharks never sleep. They swim constantly. There are two facts about their physiology that make this insomniac behavior necessary. Why must these sharks stay awake and keep on the go?

15 Fish travel long distances in the sea. Among the marathon swimming champions is the tuna fish. One particular tuna was tagged by researchers in the sea near Florida. The tagged tuna was later caught in Norwegian waters. How long do you think it took the tuna to travel the 4,200 nautical miles to the place it was caught: (a) twenty days, (b) fifty days, or (c) eighty days?

16 Do most species of ocean fish live in deep water or shallow water?

17 How many species of fish are there: (a) 12,000, (b) 24,000, or (c) 54,000?

18 You've heard of flying fish. Do flying fish really fly?

19 A certain denizen of the sea has the largest eye known to zoologists. It's also got a lot of arms. Is it a starfish, a jellyfish, or a squid?

20 Algae are fascinating organisms. If you've ever seen green scum on a pond, you've seen one kind of algae—a freshwater variety. There's another kind of algae found in the ocean. It grows (among other places) off the Pacific Coast of the United States and its strands can grow to be 200 feet long. Can you name this algae?

21 This creature, related to crabs and lobsters, is potentially the most abundant source of animal protein on Earth. It's found in Antarctic waters and is the main source of food for the baleen whale. Can you name this creature?

22 There's a sea creature that qualifies as the slowest-growing creature on our planet. Is it a turtle, a fish, or a mollusk?

23 Which sea creature makes new land where none existed before?

24 The blue whale is the largest animal ever to live on Earth. How much does a blue whale weigh compared to a full-grown African elephant? As much as ten elephants? twenty elephants? thirty elephants?

25 A certain creature living in the sea is known to live at least 200 years, making it the longest-lived animal on our planet. This oldster is (a) a turtle, (b) a swordfish, or (c) a whale?

26 There are many legends about the existence of a unicorn. Scholars of these legends think they may have been inspired by rhinoceroses. But there is an inhabitant of the ocean that also has a single "horn." It lives in Arctic waters. What is it?

27 What destructive sea creature is only a quarter-inch long, hardly ever moves, and costs ship and boat owners billions of dollars a year?

28 What's a sand dollar?

29 On one of their voyages, Christopher Columbus and his sailors reported seeing a mermaid. What did they really see?

30 In Melville's story *Moby Dick*, a crazed captain named Ahab hunts the great white whale that cost him a leg years before. Was there a real whale named Moby Dick?

31 Plankton are found in all oceans. What are plankton: (a) boards that float in the sea, (b) small animals, or (c) small plants?

32 There are more than 1,000,000 insect species on our planet. Or at least that's the number so far catalogued and given names by scientists. (It is believed that millions of unknown insect species may await discovery.) By latest exact count, how many kinds of insects make their home at sea?

33 Some species of snakes have adapted to life at sea. Full grown sea snakes are only about three or four feet long, yet again and again observers who have seen sea snakes tell wild tales of gigantic monsters of the sea. Why do people make this mistake?

34 Many minerals—gold, silver, and platinum, to name a few—are known to exist in rock formations below the surface of the sea. Today's marine geologists are making plans to exploit these resources. Measured in dollars, what is the most valuable commodity presently extracted from land beneath the sea? (Hint: it's none of those already mentioned in this question.)

35 About 9 million tons of gold have been wrested from the Earth by miners. How does that compare with the amount of gold suspended in the water of the world's oceans?

36 How much salt is there in the world's oceans?

37 An American pioneer of deep-water research, William Beebe, descended to the then-record depth of 3,028 feet off the coast of Bermuda in 1934. What is the proper name for the very strong, hollow round ball in which he made his descent?

38 The deepest descent ever made by humans into the ocean depths was achieved by Dr. Jacques Piccard and Lt. Donald Walsh in the U.S. Navy bathyscaphe *Trieste,* in 1960. How far beneath the surface did these intrepid explorers go?

39 Aboard the ship *Fram,* the Norwegian explorer Fridtjof Nansen embarked on an Arctic voyage in 1893. He purposely allowed his ship to become surrounded by polar ice, and for two years it drifted with the icepack. Among the other scientific findings of this expedition was the solution to a debate about the physical geography of our planet's far northern regions. What did the voyage of the *Fram* prove?

40 For three-and-one-half years, beginning in 1871, this British ship sailed on the first scientific expedition designed to learn about the ocean depths. Members of the expedition charted sea bottoms, studied living creatures below the surface, and mapped underwater currents. What was the ship's name? (Hint: It had the same name as an ill-fated American spacecraft that would break up shortly after launch more than 100 years later.)

41 The British Admiralty dispatched this explorer/scientist to chart the vast expanses of the Pacific Ocean in 1768. He died as the result of an arrow wound received in the Sandwich Islands (now known as Hawaii) in 1779. What was his name?

42 This pioneering American oceanographer was the first scientist to describe the worldwide distribution of winds and ocean currents in a book first published in 1847. Can you name him? Not many people can, yet his work began the modern era of knowledgeable navigation.

Answers

① *You'd never have time to count them all, for there are about 1,700,000,000,000,000,000 (that's 1.7 quintillion) water molecules in a single drop of sea water.*

② *In the Bay of Fundy on the coast of Nova Scotia, Canada, where the customary difference between water levels at low and high tide is fifty-seven feet.*

③ *Ebb tide is low tide.*

④ *The correct answer is (c). A tsunami is actually a huge wave set in motion by an undersea earthquake. Tsunamis are often (incorrectly) called "tidal waves."*

⑤ *Faster. Sound travels at an average speed of about 3,350 miles per hour in water—almost more than four times faster than it does in the atmosphere.*

(6) Before you give up, think of the inventor of bifocal glasses, a more efficient wood-burning stove, or an investigator of lightning. Right. The man who first mapped the Gulf Stream was none other than Benjamin Franklin.

(7) The ships had encountered the Gulf Stream.

(8) Canada wins by a big margin, with more than 56,000 miles of coastline. Indonesia is second, with 34,000 miles of ocean shore.

(9) Aristotle was reputedly proud of his ability to explain anything. The story goes that he was so disturbed by his inability to explain the mystery current that he committed suicide. Incidentally, oceanographers are still puzzled by this current, but they don't take their puzzlement as hard as Aristotle allegedly did.

(10) The messages were simple numbers indicating the ships' position and date. When the bottles were picked up along the coast, the finders sent them to the U.S. Navy's Department of Charts and Instruments. The place and time they were picked up helped reveal the speed and direction of ocean currents.

(11) The answer is an amazing ten hours. Why so long? You know that pressure and density increase with depth. The deeper the stone gets,

the more resistance it meets, so it sinks more and more slowly the farther it falls.

⑫ *For safety reasons. They don't want the 30-pound scuba tanks falling on top of them.*

⑬ *Most limestone forms from calcium carbonate precipitated from rivers as well as shells and coral skeletons that collect on the bottom.*

⑭ *First, most sharks must constantly swim to keep oxygen-rich water flowing past their gill slits. Second, unlike most other ocean fishes, these sharks lack air bladders that enable fish to float.*

⑮ *The tagged tuna made the 4,200-mile run in fifty days, therefore (b) is the correct answer.*

⑯ *Most ocean fish live in shallow water. Eighty-five percent of all ocean fish live in the shallow waters over continental shelves or above reefs.*

⑰ *The correct answer is (b). There are about 24,000 known species, but scientists think there may be a lot more still to be discovered.*

⑱ *This question might start an argument at a convention of marine biologists. Some observers swear that they have seen flying fish actually flapping their "wings" (which are really fins) as they course through the air, but slow-motion films of flying fish show them gliding, not really flying. They leap out of the water to avoid predators (such as tuna) and can rise as much as forty feet in the air, traveling more than the length of two football fields before re-entering the water.*

⑲ *It's a squid, more specifically the giant squid. A relative of octopuses, this squid has eyes as large as fifteen inches in diameter. How would you like to come eyeball-to-eyeball with one of those?*

⑳ *It is called kelp. It attaches to rocks on the seabed and grows upward toward sunlight, forming huge underwater "forests" that provide a home to many marine animals.*

㉑ *It's a shrimp-like creature (scientific name:* Euphausia superba; *English name: krill). Adult krill are about two-and-one-half inches long. They swim in huge schools. There may be more than 5 billion tons of krill swimming in frigid southern waters.*

㉒ *It's a mollusk, a species of deep-sea clam that takes an entire century to reach its full-grown size of less than one-third of an inch.*

㉓ *The corals. Living corals attach themselves to the dead skeletons of their kind, thus building up coral reefs and islands known as atolls.*

(24) *If you said thirty elephants, you're right. Adult blue whales can weigh as much as 200 tons.*

(25) *The correct answer is (a). A 200-year-old sea turtle is a matter of record, but scientists believe that some turtles may live even longer.*

(26) *It's the narwhal, a species of whale. The "horn" is not really a horn, but a tooth that can grow as long as nine feet.*

(27) *The answer is the barnacle, a small, shelled creature that emits a super glue that fastens it to rocks, sand, other creatures, or to the hulls of ships and boats. Scraping barnacles off boats or killing them with chemicals costs billions of dollars a year. Small boat owners experience lots of hard work and exasperation as they try to remove the barnacles that slow the passage of their boats through the water.*

(28) *A sand dollar is not made of sand. It is a creature, related to the sea star. It burrows in the sand to avoid being eaten by fish like the haddock, which enjoy munching on its kind.*

(29) *A manatee, or sea cow—a creature that certainly doesn't look much like a mermaid. This seagoing mammal can be more than ten feet long and weigh half a ton. Manatees (which are still found in Florida) were not known to Europeans. Maybe nobody aboard Columbus's ships got a really good look at one.*

(30) Not that anybody knows of. Melville did, however, base much of his story on a real whale that whalers had named Mocha Dick, a sperm whale that killed many men who tried to hunt it. In 1859 when Mocha Dick was finally killed, he was found to have more than twenty harpoons imbedded in his body.

(31) Both (b) and (c) are correct answers. The word plankton means "drifting" and plankton are the countless microscopic plants and animals that float in the sea. Though small, they are very important to ocean ecology and provide food for many larger creatures.

(32) Only one! Its scientific name is Halobates. In English it is called the sea strider, or more colloquially, the "Jesus bug." The latter name is given this insect because it walks on water, supported by surface tension.

(33) Because sea snakes swim in long processions. Thousands upon thousands have been observed swimming in strips several feet wide and many miles long. Thinking this mass of snakes is one creature, many people who see it report seeing a sea monster.

(34) It is petroleum (usually called oil) that is pumped from geological formations beneath the sea by many offshore oil pumping platforms all around the world.

(35) *The 9 million tons dug from the Earth by miners almost exactly matches the amount estimated to be suspended in ocean water.*

(36) *It is estimated that the oceans contain a staggering 50,000,000,000,000,000 (that's 50 quadrillion) tons of salt. If all of this salt could be spread out evenly over a globe the size of the Earth, it would form a layer 500 feet thick.*

(37) *Beebe's deep-sea device was called a bathysphere.*

(38) *These explorers descended 35,280 feet—almost seven miles. At that depth the* Trieste *was subjected to a crushing pressure of more than eight tons per square inch.*

(39) *Nansen's observations proved that the regions around the North Pole were covered by deep ocean waters. Others had thought that there might be land beneath the polar ice.*

(40) *The British ship was the* H.M.S. Challenger.

(41) *This eighteenth-century explorer/scientist was Captain James Cook.*

(42) *He was Matthew Fontaine Maury, of the U.S. Navy. He compiled his pioneering navigational observations from the logbooks of Navy vessels.*

Secrets in Stone

What is the Earth made of? How are mountains made? What causes volcanos to spew their molten rock? What forces account for earthquakes? Where can vital resources such as metals and petroleum be found? Scientists who seek answers to these questions are called geologists.

How much do you know about the Earth we live on and the processes that shape it? Let's find out.

❶ Geology, properly speaking, is the study of the whole Earth. But to many of us the word means "study of rocks," and, indeed, the study of rocks is an important aspect of earth science. Do you know who first wrote down what was known about rocks?

❷ Traditionally, geology is divided into two broad fields. What are they?

❸ How do geologists use measurements of the radioactivity in rocks to determine the age of the rocks?

❹ The oldest rocks ever found on Earth are dated by geologists at about 4.2 billion years old. That's "only" a few hundred million years younger than the planet. Where were these oldest rocks found.

❺ Let's revive an old argument. Is the Earth round?

❻ Have you ever found a fossil? If you have, most likely it was the fossil of an ancient creature of the sea. Fossils of sea creatures are found all over the world, in locations that are today far from any ocean. Who was the first person on record to conclude that this meant that once upon a time many areas that are now dry land were once beneath the sea?

❼ Here's a basic question about the Earth. Is it getting heavier or lighter with the passage of time?

❽ Scientists have come up with a figure for what they believe is the temperature at the center of the Earth. They've also calculated the temperature of the surface of the sun. Which is believed to be hotter?

❾ Study of the Earth has revealed that its crust has been torn and twisted by awesome forces. What geological event created "the loudest sound ever heard"?

10 In 1943 a farmer in Mexico was surprised to see a hole in one of his cornfields. The hole was spitting hot cinders into the air. The farmer tried to cover the hole with dirt, but every day it kept getting bigger, and within weeks a baby volcano was growing in the cornfield. What is the name of the Mexican volcano that grew from nothing?

11 Gravitational attraction by the sun and moon causes the ocean tides that ceaselessly sweep the seas. Do tides affect the land as well as the sea?

12 You know that Earth is like a giant magnet, with a North Magnetic Pole and a South Magnetic Pole. You may also know that magnetism is a mystery to scientists (nobody knows how it works). But geologists do have some ideas about what forces produce Earth's magnetic field. Is Earth's magnetism produced by (a) radioactivity, or (b) currents of molten rock that make electricity?

13 You may know that the geographical North Pole and the North Magnetic Pole are not in the same location. Approximately how far apart are they: (a) 200 miles, (b) 600 miles, or (c) 1,200 miles?

14 At the North Pole, vast ice fields float on the Arctic Ocean. At the South Pole, ice sheets blanket Antarctica. Is there more ice at the North Pole or the South Pole?

15 Glaciers form high on mountains and near Earth's poles. Where does the ice in glaciers come from?

16 Where is the world's longest glacier?

17 Was the Sahara Desert ever covered with ice?

18 Born in Switzerland, Jean Louis Rodolphe Agassiz studied glaciers. His research indicated that at one time huge ice sheets had covered much of Europe. When Agassiz moved to the United States, he helped to prove scientifically that parts of the United States too had also once lain beneath thick blankets of ice. One of the largest lakes on Earth was named for him. Where is it?

19 Where did the rocks of the continents come from? This question has always fascinated geologists. Can you name the Scottish physician who was the first scientist to recognize that many rock layers must have been produced by volcanic action?

20 Another Scottish scientist proved that continental rocks could be formed by heat. He melted rocks in huge furnaces. He found that melted limestone turned into marble, and that rock

similar to granite formed when volcanic rocks were melted and allowed to cool. Can you name this geologist?

21 In 1912 a German scientist, Alfred Wegener, proposed what seemed like a crazy theory. The continents, Wegener said, were moving around, floating on the molten interior of the planet. Most geologists pooh-poohed Wegener's ideas. Continents were just too big to move, they said. Who was right?

22 How did studies of the magnetism in rocks help prove that pieces of the Earth's crust move around in relationship to each other?

23 Once upon a time (about 300 million years ago), all the world's continents were united into one supercontinent. What name have geologists given this landmass of yesteryear?

24 Are New York City and London closer together or farther apart than they were 100 years ago?

25 To human eyes, mountains are huge structures, and the very highest of them are certainly hard to climb. What is the highest mountain on Earth?

26 You've seen pictures of Mt. St. Helens blowing its top in May of 1980. The

volcanic explosion hurled ash twelve miles straight up into the air. A shock wave flattened trees for miles from the mountain's summit. But did you know that Mt. St. Helens prevented ocean freighters from sailing, even though the volcano is eighty miles from the ocean. What effect did the volcano have that obstructed shipping?

27 Earthquakes, produced by the sudden shifting of pieces of the Earth's crust, send out shock waves that travel through and around the world. Today scientists use sensitive instruments to measure and record these shock waves. What are these instruments called?

28 About how many earthquakes strong enough to damage buildings or cause injury occur in one year (on the average)?

29 What town in Alaska was rebuilt in a new location after it was destroyed in a 1964 earthquake?

30 Geologists call them "smokers." Are they found at sea or on the land?

31 "The girls can flirt and other queer things can do." What do these words mean to geologists?

32 A diamond is the hardest substance known. If that is so (and it is), what substance do diamond cutters use to cut diamonds?

(33) You probably know that diamonds are a form of pure carbon squeezed together deep in the Earth's crust under conditions of high temperature and pressure. And Superman can make diamonds by squeezing coal in his bare hands—right? Ordinary mortals can make diamonds too. In 1955 scientists made the first man-made diamond by producing pressures of more than 100,000 atmospheres and temperatures of more than 2,500 degrees Centigrade. Question: Are man-made diamonds as hard as natural diamonds?

(34) If you put a diamond in a hot flame, will it stay unchanged, get soft, or burn up?

(35) Diamonds are found in only one place in the United States. Where?

(36) What do diamonds have to do with volcanos?

(37) What gemstone is the second-hardest mineral, and what color is this gem?

(38) Most precious stones are transparent—just think of diamonds, rubies, sapphires, and emeralds. There is one gemstone, however, that is opaque. Which one?

(39) The long and complicated geological history of our planet has created thousands of different kinds of mineral crystals, many of them quite beautiful and therefore prized by collectors. Amethyst crystals are sometimes found in disguise.

They hide inside hollow rocks that show no sign of their hidden treasures. What are these hollow rocks that contain crystals called?

40 Sand, one of Earth's commonest minerals, is formed by weathering processes that break down rocks. How can geologists tell old grains of sand from grains formed more recently?

41 Gold! Human greed for gold has made it one of the most-searched-for minerals on Earth. Besides having eye appeal, gold has other properties that make it so valuable. Gold's "ductility" is one of its most amazing properties. Ductility measures how well a metal may be drawn out into a thread. How long a thread can be made from one ounce of gold?

42 "All that glitters is not gold." Many people have thought they were rich when they discovered glittering veins of a mineral called "fool's gold." Fool's gold is actually iron pyrite. Its crystals are pretty, but hardly valuable enough to make anyone's fortune. Iron pyrite, however, is not totally worthless. It is used to produce one of modern technology's basic chemicals. What is this chemical?

43 An incredible amount of geological knowledge has been gathered in mankind's search for petroleum—the "juice" that makes our world run. Oil wells are by far the deepest "mines" ever dug in search of mineral riches. How deep are the deepest oil wells?

Answers

① This pioneering geologist, Theophrastus, was a student of the great Greek philosopher, Aristotle. His paper, called "Concerning Stones," is the first known attempt to record all that was known about rocks, minerals, and fossils.

② They are physical geology and historical geology. Physical geology is the study of the Earth's materials and the forces that shape the Earth. Historical geology attempts to understand the history of the Earth.

③ All rocks contain traces of radioactive elements. These elements decay at a known rate into other elements. For instance, uranium-235 decays into lead-207 at a known rate. By measuring how much of these two elements are in a rock sample, geologists can tell the age of the rock.

④ The oldest rocks ever found on Earth were discovered in the desert of western Australia in 1983.

⑤ Once upon a time, so the story goes, people thought the world was flat, and then scientists proved it was round. Actually, the Earth is only "almost" round. It has bulges and bumps. It bulges out around the Equator, for instance, because of Earth's spin.

⑥ Herodotus, who traveled widely in Mediterranean lands in the fifth century B.C., *was the first to write that shells of sea creatures found on land proved that the land was once underwater. Nobody believed him for centuries, but he was right.*

⑦ *It's getting heavier, scientists say. Everv day it is estimated that 100 tons of meteoritic material enter the Earth's atmosphere. Even though most of the meteors "burn up," their dust eventually settles to Earth.*

⑧ *Maybe you thought this was a trick question and answered that the center of the Earth is hotter. If so, you're wrong, but not by much. Scientists believe that temperatures at the center of Earth are about 9,000 degrees (Fahrenheit). The surface of the sun is believed to be cooking away at a temperature just somewhat above 10,000 degrees (Fahrenheit).*

⑨ *As far as we know, it was the volcanic eruption of the island of Krakatoa. This awesome series of explosions blasted nearly ten cubic kilometers of rock into the atmosphere. Ash from the eruptions spread around the world. Waves from the vibrations surged through the world's oceans. The largest single explosion occurred shortly after ten o'clock on the morning of August 27, 1883. The sound was heard almost four hours later on tiny Rodrigues Island in the Indian Ocean, nearly 3,000 miles away. No other sound has ever been reported as having traveled so far. Thus the eruption of Krakatoa qualifies as the loudest sound ever heard.*

⑩ *The volcano is named Paricutin. It continued to erupt for ten years and then became quiet again.*

⑪ *Well, the continents don't move up and down as much as the ocean, but they do rise or fall as much as six inches under the influence of the sun and moon.*

⑫ *Both the answers are correct. Radioactive heat deep in the Earth keeps rock in the outer core molten. Huge flows of this molten rock generate electrical currents that produce a magnetic field.*

⑬ *The correct answer is (c). It surprises most people to find out they are 1,200 miles apart.*

⑭ *It's not even a contest. Nine-tenths of the ice on Earth is clumped around the South Pole in the great Antarctic ice sheet, where the ice has built up over the centuries to a depth of nearly three miles at its thickest.*

⑮ *The ice that builds glaciers is formed from snow. When snow falls and does not melt it builds up into ever-deeper layers. Gradually, snow that falls on glaciers is squeezed into ice by the weight of new snow falling on it. How long does this take? That depends. . . in mountains of Europe or North America, it may take only five years for snow to become ice. In northern Greenland or Antartica, where temperatures are much colder, and snowfall less, snow may fall and lie atop the underlying ice for thousands of years before more snow falling squeezes it down and turns it into ice.*

(16) *If you said somewhere on Antarctica, the continent of ice, you were correct. Antarctica's Lambert Glacier and Fisher Glacier system is more than 300 miles long and more than forty miles wide at its widest.*

(17) *Yes. Geologists studying plate tectonics (the movement of the Earth's crust) believe that about 500 million years ago, the section of the Earth's crust that now forms the Sahara Desert was near the South Pole and was covered with thick sheets of ice.*

(18) *Lake Agassiz no longer exists. At one time it lay at the edge of the ice sheet that covered what is now Canada and the northern United States. Lake Agassiz stretched across much of what is now the states of Minnesota and North Dakota, and the province of Manitoba.*

(19) *The physician/geologist was James Hutton. Hutton was also the first geologist to recognize that the processes that shaped the Earth in its past were much the same as those going on in the present. When his book explaining these theories was*

published in 1785, few other scientists accepted these theories. Today Hutton's ideas are considered to be the foundation of modern geological science.

⑳ *He was Sir James Hall.*

㉑ *Wegener was right. The continents are moving. The study of their movement is called plate tectonics.*

㉒ *Scientists discovered that different rocks in different parts of the world showed different magnetic alignments. This showed that they had been formed when the continents occupied different positions on the surface of the Earth.*

㉓ *The supercontinent of yesteryear is called "Pangea" by geologists.*

㉔ *They're farther apart. North America and Europe are moving away from each other at the rate of about one and one-fifth inches a year. Thus, in 100 years, New York and London have drifted apart by about ten feet.*

㉕ *There are two correct answers to this question. Mt. Everest, in the Himalayas, measures 29,000 feet above sea level (give or take a few feet), thus qualifying as the world's loftiest peak. But Mauna Kea, in the Hawaiian Islands, is a volcano that reaches 14,000 feet above sea level and also goes down to the sea bottom 19,000 feet below the sea surface. So from its base to its summit, Mauna Kea measures more than 33,000 feet, thus making it the world's "tallest" mountain.*

㉖ *Before Mt. St. Helens blew up, it was covered with ice and snow. The volcanic eruption melted this and created huge surges of boiling mud that swept down the sides of the mountain. So much mud flowed that it clogged shipping lanes in the Columbia River.*

㉗ *They are called seismometers. The* seismo *part of the word is taken from the Greek word for earthquake, and the* meter *part means measure.*

㉘ *Every year there are about 1,000 earthquakes that cause damage. One hundred times as many earthquakes occur that are felt by people, and at least half a million earthquakes a year are recorded by sensitive seismographic instruments.*

㉙ *The town was Valdez, which is located a little more than 100 miles from Anchorage on the shores of Prince William Sound. During the great Alaskan earthquake, Valdez was hit by a double whammy. First the ground beneath rolled in huge ripples as the shock waves from the earthquake shook it. Then the town was flooded by tsunami waves (sometimes called "tidal waves") set surging by the earthquake.*

㉚ *"Smokers" are underwater hot springs in the ocean. They actually emit water darkly colored with minerals, not smoke.*

㉛ *The phrase is a mnemonic (memory device) for helping geologists remember the Mohs (rhymes with "toes") Scale, which defines the hardness of rocks and minerals. The first letters of each word indicate the minerals Talc, Gypsum, Calcite, Fluorite, Apatite, Orthoclase, Quartz, Topaz, Corundum, and Diamond. Talc is, of course, the softest mineral in the series and diamond the hardest.*

㉜ *They use other diamonds Most of the diamonds mined on Earth are used for industrial purposes such as grinding and cutting.*

㉝ *Every bit of it. Manmade diamonds are chemically identical to natural diamonds and just as hard (and, most say, just as beautiful). They are widely used for their abrasive and cutting properties.*

㉞ *Considering that a diamond (like coal) is composed of carbon, you shouldn't be too surprised to learn that diamonds will burn.*

㉟ *Near Murfreesboro, Arkansas. Though a few large diamonds have been found in the area, they are not commercially mined.*

㊱ *All natural diamonds were formed in "pipes" of molten rock inside volcanos.*

(37) *Ranking just below diamonds on the hardness scale are sapphires. What color did you say? They can be any color except red. Why not red? Because red stones similar to sapphires are called rubies.*

(38) *The opaque gemstone is opal. Most opal is mined in Australia.*

(39) *The hollow rocks containing crystals are known as geodes.*

(40) *Newly-formed grains of sand tend to have sharp corners. When sand is moved about over time by wind or water, collisions cause these corners to break off. So an old grain of sand is much rounder than a young one.*

(41) *One troy ounce of gold (which is slightly heavier than the more common avoirdupois ounce used for most other materials) can be drawn out into a wire more than thirty miles long!*

(42) *Iron pyrite is used in producing sulfuric acid, a basic chemical used in the manufacture of many products.*

(43) *About six miles. If we continue to use petroleum resources at our present rate, deep drilling may become more necessary in the future. Some geologists believe that huge reserves of oil lie deep in the Earth.*

Fossil Facts

Throughout history, people have sought knowledge of Earth's past. Much of what we know about the plants and animals that once lived on our planet we have learned from fossils, the hardened remains or traces of life left from previous geological epochs. The Greeks of 600 B.C., having found fossil shells far from ocean shores, concluded that ancient seas must once have covered parts of what by then was dry land. Bones uncovered by flood waters, or dug up by humans, gave unmistakable evidence that creatures no longer in existence had once roamed Earth's hills and valleys and lived in its rivers and oceans.

From the Greek words for "ancient life" came the word for the science we call "paleontology." You can become a "paleo-detective" as you dig into this trove of fossil facts.

1 When we think of fossils, we tend to think of fossil bones, because these hard structures survive long after softer body parts have vanished. But not all fossils are those of bones. The oldest known fossils, for instance, are those of creatures that had no bones. What are these fossils, and how old are they?

2 Life on Earth originated in the oceans. In the first 3 billion years of evolution, there were only two kinds of living creatures. What were they?

3 The algae that teemed in the oceans of long ago were the first plants. They absorbed the energy of sunlight. In the process they released a substance that eventually caused many of them to die off. Yet we think of this substance as a necessity for life. What is it?

4 How long has animal life existed on Earth?

5 Even if you don't know much about fossils, you may recognize fossils of this ancient creature when you find them. And find them you may, for they are widespread and common. Scientists aver that there were several thousand genera and perhaps 10,000 species, living from 250 million to 600 million years ago. They were among the first animals to have efficient eyes. Fixed at a variety of angles, they enabled this creature to see in many directions at once. What creature left behind this common, everyday fossil?

6 Most of the fossils found and studied by paleontologists were formed underwater. Why is this so?

7 No other fossils excite as much interest and awe as those of the fascinating creatures we call dinosaurs. What does the word *dinosaur* mean?

8 How long ago did the first dinosaurs walk the Earth?

9 It was probably the first fossilized bone of a dinosaur reported in scientific literature. It was discovered near Woodbury Creek in Gloucester County, New Jersey, and was described in a report to the American Philosophical Society in 1787. Where is this fossil now?

10 The first scientific recognition that dinosaur fossils were the remains of creatures that became extinct long ago was given in England in the 1820s. An English clergyman, William

Buckland, was the first scientist to name and describe a dinosaur—a creature forty feet long and seven feet tall, related to ancestral crocodiles. What name did Buckland give this creature?

11 *Tyrannosaurus rex* is commonly depicted as the most fearsome dinosaur. Was it really the biggest and most deadly, truly the king of the dinosaurs?

12 Fast-running mammals have shins that are long in comparison to their thighs. Take the length of an animal's shin, plus the length of the ankle. Divide that by the length of its thigh. If you come out with a number that is 1.5 or higher, you know the animal was a moderately fast runner. By measuring the shins and thighs of dinosaurs, we have learned that dinosaurs were not all slow and lumbering, as they are usually depicted. Some had to have been graceful and swift. How does the speed of the fastest dinosaur compare with the speed of a fast-running animal of today?

13 In a sandstone quarry in Scotland, Dr. Alick Walker found no fossil bones but he did observe a number of oddly-shaped holes in the rock walls the quarry workers had exposed. From these holes the paleontologist was able to reconstruct the complete skeleton of a dinosaur. How did he do it?

14 In 1922 Roy Chapman Andrews, of the American Museum of Natural History, led a scientific expedition into the remote Gobi desert of central Asia. This is a region that had never been covered by the seas, so the explorer thought his twenty-six-man team might discover fossils of previously unknown mammals. They didn't find many remains of what they were looking for, but they made a spectacular discovery about dinosaurs. What did they find that excited the world of science and settled a controversy?

15 For nearly 100 million years of their development, the most complex land-dwelling animals were amphibians, which meant they had to spend some of their time in water. Like their modern descendants, the original amphibians laid their eggs in water. What evolutionary development was necessary before animals could lay their eggs on land?

16 What did the first animals capable of spending their entire lives on land look like?

17 What are the oldest land-dwelling animal fossils yet found?

18 In 1737 the Swedish naturalist Linnaeus gave the name *Reptilia* to creatures such as lizards, turtles, crocodiles, and snakes. Why did he choose this name?

19 The heaviest bird that ever lived was *Aepyornis maximus* ("greatest of the high birds"). It weighed close to half a ton. Another record breaker was *Dinornis giganteus* ("giant terrible bird"), the tallest known bird. It stood more than eleven feet high. What do these two huge birds have in common?

20 One day in 1938 a fisherman off the eastern coast of South Africa landed a large fish that puzzled him. The fisherman didn't know what it was, but he was sure it was a kind of fish he had never seen before. When he took it to a museum, the curator made an announcement that startled the world of science. It was a strange fish indeed. What was so remarkable about it?

21 One of the largest invertebrates that ever lived was *Eurypterides*. This invertebrate inhabited our planet during the Silurian period, 440 to 395 million years ago. You definitely would not want to come across one of these creatures. Why?

22 Amazingly, a creature of a type still alive on Earth today was the terror of the dinosaurs. Do you know what it is?

23 Were this animal alive today, it would be browsing on leaves high in the treetops. It was the largest known land mammal. Can you name it?

24 How long ago did the first insects evolve?

25 The largest insects ever to live on Earth were giant dragonflies which, 400 million years ago, flew through forests of huge mosses and the primitive ancestors of modern trees. Did their wings span (a) two feet, (b) three feet, or (c) five feet?

26 Sometimes the nature of an extinct creature can be inferred from very fragmentary evidence. Gideon Mantell, a British physician fascinated by fossils, believed that some very large fossil teeth found in rock quarries of the Tilgate Forest region of England had belonged to a large, extinct reptile. Other scientists came to agree with him after Mantell found that remarkably similar, though much smaller, teeth were found in a modern reptile of Central America. Mantell named his extinct creature after the living reptile. What did he call it?

27 In 1977 Russian paleontologists made an astonishing find in Siberia. What they discovered provided important genetic information about today's elephants. What was it that they retrieved from the frozen earth?

28 Fossil hunters have found the remains of many ancestors of present-day elephants. How many different kinds of elephants do you think there have been on Earth: (a) 50, (b) 100, or (c) 600.

29 Why do searchers find so many shark's teeth?

30 Were penguins around in prehistoric times?

31 Most of the fossils we find tell us that the ancestors of today's animals were larger than their modern-day relatives. The biggest animal on Earth today is the blue whale, which can be over 100 feet long and weigh more than 400,000 pounds. Were its long-ago ancestors even bigger?

32 A creature named *Hyracotherium* roamed the forests of North America and Europe in the late Palaeocene and early Eocene periods. Fossil remains have told us it was Earth's first horse, the ancestor of all horses. Was it larger or smaller than horses of today?

33 The largest bird capable of flight was the *Argentavis,* which flew around in what is now Argentina in the early Pliocene period (between 2 and 7 million years ago). *Argentavis* looked something like present-day vultures. Want to guess at its wing span? Was it ten, fifteen, or twenty-five feet?

34 How can scientists tell from examining fossil teeth whether the animal from which they came ate leaves or grass?

35 More than 100 million years ago resin that leaked from trees sometimes imprisoned ants and other insects before it hardened. Creatures thus trapped can still be seen today. What is the clear substance in which they are perfectly preserved?

36 Edward Hitchcock, a nineteenth century theologian and scientist, devoted many years to the collection of a somewhat unusual kind of fossil. Eventually, his collection required an entire floor of a museum at Amherst College, where he was a professor. His findings were not bones. What are the objects in this famous collection?

37 In 1810 an eleven-year-old English girl named Mary Anning found a strange fossil. She helped to dig it up and was credited with discovering the first skeleton of a prehistoric sea creature, *Ichthyosaurus*. Later, she made other finds, including that of *Dimorphodon*, one of the first flying reptile fossils ever discovered. Why did she continue to look for fossils for the rest of her life?

38 Of all the species that have ever lived on Earth, what percentage is still alive today?

39 What do we call the phenomenon that occurred, somewhere between 280 and 225 million years ago, when almost all of the species then living on Earth became extinct?

40 Has science found a satisfactory explanation for the Permian extinction?

41 Paleontologists studying reefs, where corals lay down a layer of their skeletons each day, learned a lot more than facts about these strange creatures of the sea. Through careful examination of the layered deposits, scientists came to an understanding of their relationship to lunar cycles. What amazing fact did these fossil corals reveal about Earth?

Answers

① *The oldest fossils are those of microscopic, one-celled creatures. They are found in rocks formed about 3.5 billion years ago. These fossils show up as patterns of mineralization, and are, of course, visible only with the aid of a microscope.*

② *The two kinds of living creatures were bacteria and algae.*

③ *It is oxygen, which was toxic to the life-forms that produced it.*

④ *Although the fossil record is unclear and cannot be considered complete, scientists believe that animals—forms capable of purposeful movement—came into existence about a billion years ago.*

⑤ *The familiar trilobite.*

⑥ *The remains of creatures who died and sank to the bottoms of rivers, lakes, and oceans are often covered with mud or sand that kept them from decaying completely.*

⑦ *The word* dinosaur *was coined from two Greek words meaning "terrible" and "lizard." Thus dinosaur means "terrible lizard."*

⑧ *Paleontologists believe that the first dinosaurs evolved during a geological era called the Triassic, about 200 million years ago.*

⑨ *Nobody knows. It has disappeared. From its description, paleontologists think it may have been a bone of a dinosaur called a duck-billed* Hadrosaur.

⑩ *He called it* Megalosaurus, *which means "giant lizard."*

⑪ *For a long time scientists thought so, but paleontologists digging near Fort Collins, Colorado, found fossils of a giant dinosaur that probably out-weighed and out-killed the mighty* Tyrannosaurus rex. Epanterias amplexus *is the name of this monster, which roamed our planet about 130 million years ago during the Jurassic period.*

⑫ *Amazingly, some dinosaurs had a shin-to-thigh ratio of two or more, matching that of present-day gazelles.*

⑬ *The scientist knew the holes had been left by bones that had dissolved. He used the holes as molds, filling them with a soft plastic. When he removed the hardened plastic, it had the shape of the bones that had once occupied the holes.*

⑭ *The Andrews expedition discovered nests of dinosaur eggs. Up to that time, scientists were not sure that dinosaurs laid eggs.*

⑮ *It is called the amnion. The first amnions were soft, flexible sacs that contained a fertilized egg and a supply of moisture in which a young animal could live while it grew into its adult form.*

⑯ *They looked very much like present-day lizards. They were the first members of the class of creatures we call reptiles.*

⑰ *They are arthropods, which looked a lot like modern scorpions. Their name means "having jointed legs." The primitive arthropods, which became the first land-dwelling animals, crawled out of the sea about 415 million years ago.*

⑱ Reptilia *comes from a Latin word that means "creeping on the belly" or "crawling."*

⑲ *They are two prehistoric creatures that managed to survive until about 1700 A.D. The first lived in Madagascar, the second in New Zealand.*

⑳ *The fish was a living fossil. It was a coelacanth, a kind of fish that had been believed extinct for 60 million years. Scientists had seen many coelacanth skeletons, but never a live fish. Since then, more living coelacanths have been found in deep water off Madagascar.*

㉑ *Because it was a giant scorpion-like creature, six feet long, with huge predatory pincers and a poisonous sting.*

㉒ *It's the crocodile. In the time of the dinosaurs, the giant* Deinosuchus, *fifty-two feet long, ambushed dinosaurs when they came to rivers to drink, easily crushing them in its fearsome jaws.*

㉓ *It was* Paraceratherium, *a giant, hornless rhinoceros, eighteen feet in height at its shoulders.*

㉔ *The oldest insect fossils are 413 million years old.*

㉕ *The right answer is (b). These giant flying insects measured slightly more than three feet from wing tip to wing tip.*

㉖ *He christened it* Iguanodon, *meaning "iguana tooth." Like most dinosaurs,* Iguanodon *was a plant eater. Almost thirty feet long, it lived in swampy lowlands.*

㉗ *They dug up a completely and perfectly preserved young mammoth. Analysis of its blood and tissues further established the relationship of this primitive pachyderm to the elephants of today.*

㉘ *There have been a lot of different kinds of elephants. The correct answer is (c) 600.*

(29) *Many parts of the world were covered with oceans and there were lots of sharks, which lost lots of teeth. Young sharks replace each tooth about once a week.*

(30) *Were they ever! Penguins of the Eocene period (38 to 54 million years ago) were giants, standing over five feet high.*

(31) *No. They were much smaller, ranging in size from seven to thirteen feet in length.*

(32) *Much smaller. This primitive horse was only about the size of a fox.*

(33) *The* Argentavis *had a wing span of twenty-five feet.*

(34) *By how much, and in what way, the teeth are worn down. Grass is usually coated with dust, dirt, and silica, which are abrasive. A grass-eater's teeth have a short life. Since leaves are cleaner and contain no silica, the teeth of leaf-eating animals last longer.*

(35) *The substance is amber.*

(36) *Hitchcock's collection consists of the fossilized footprints of dinosaurs.*

(37) *Mary Anning, who never became a scientist, had one major reason for her skillful search for fossils. She did it to make money. She sold them for good prices to collectors.*

(38) *Only 1 percent.*

(39) *This phenomenon is known as the Permian extinction.*

(40) *One of the most likely explanations is based on what is called "the iridium anomaly." In many places in the world, layers of iridium, an element not of Earth origin, have been found. The strata in which these iridium layers are located were deposited at the time of the Permian extinction, leading to speculation that the iridium might have been brought to our planet by a giant meteorite that caused worldwide destruction.*

(41) *Scientists came to the surprising conclusion that Earth's lunar month was once longer than it is now. Perhaps even more astonishing was the realization that Earth's year was once 428 days long.*

People of the Past

Human-like creatures have walked the continents of Earth for only a few million years. What do we know about our earliest ancestors? How did they live? Where did they live? These questions and many more occupy the minds of paleontologists who study fossil bones and ancient dwelling sites.

Slowly, our ancestors developed superior technologies that eventually led to the development of civilization. Archaeologists, who dig up the ruins of ancient cities, attempt to understand the workings of complex human societies of long ago.

The questions that follow will probe

your knowledge of human life before the development of the written record we call history. How many can you answer?

❶ There is still scientific argument, but most researchers are in agreement about the continent on which human beings developed. Can you name it?

❷ What may be the oldest human bones on Earth were found in 1974, in the searing deserts of northern Ethiopia. The bones unearthed by anthropologists Don Johanson and Tom Gray were those of a female, human-like creature who lived more than three, but less than four, million years ago. Whether or not this creature was a direct ancestor to people or not, she has been given a human name by scientists. What do they call her?

❸ Nearly 4 million years ago, at least two creatures who may have been ancestral to or similar to humans walked upright on two feet near an active volcano in what is now East Africa. How can we be sure just how these creatures walked so long ago?

❹ For millions of years, our human ancestors lived a kind of hand-to-mouth existence. They hunted animals and ate wild plants.

What do scientists call this stage of human development?

5 Humans are not the only creatures to use tools. Otters use rocks to crack open shellfish. Birds use thorns to pry open fruits. Apes and monkeys use sticks and stones for a variety of purposes. And there is no doubt that our ancestors used objects they found as tools. But, so far as we know, humans are the only creatures who make tools. The oldest evidence of toolmaking was discovered on the shores of Lake Turkana, in East Africa. Shaped stone choppers and hammers used for butchering were found in a layer of volcanic ash that can be dated very accurately. What is the estimated age of these tools?

6 One of the most important discoveries made by our distant ancestors was how to make fire and use it for cooking, for heat, and in the manufacture of tools and utensils. Scientists have learned a lot about the life of early humans by analyzing the remains of fires. The oldest solid evidence of the human use of fire was found near Lake Baringo, in Kenya, Africa, in 1981. How long ago do we believe those fires burned?

7 The primary method used in archaeological investigation is digging. But when modern archaeologists excavate a site, they don't dig it all up; they leave strips of undisturbed earth between the holes they excavate. What are these strips called? Why are they left?

8 For a long time the archaeological evidence seemed to show that the making of metal tools began about 3,500 years ago in what is now Iraq, where people made tools, weapons, and ornaments of bronze. But, in 1974, a discovery was made in southeast Asia that seemed to indicate a knowledge of working in bronze that predates the Middle Eastern finds. In what country was this evidence of human use of bronze discovered?

9 How long ago did the first humans live in the lands we now call Europe?

10 The human ancestors who first moved to what is now Europe had to adapt to a colder climate. Archaeological research in France has unearthed an ancient tool that proves that the people we call Cro-Magnons had invented sewn clothing at least 30,000 years ago. What humble artifacts prove this to be true?

11 Undoubtedly one of the most important revolutions in the life of humans was the development of agriculture, which made it possible for people to live in settled communities and eventually led to the development of cities. Among the first farmers were a people who lived nearly 9,000 years ago in what is now southern Turkey. These farmers cultivated two seed bearing grasses, descendants of which are still important grain crops. What were these two grains?

⓬ Scientists believe that stories of the Garden of Eden came from long ago civilizations in Mesopotamia. The word *Mesopotamia* means "between the rivers." What are these two rivers?

⓭ In the mid-part of the nineteenth century, archaeologists excavating at the ruins of the Assyrian city of Nineveh discovered clay tablets that tell a tale believed to be the forerunner of the Biblical story of the flood. Like Noah, Utnapishtim, a character in this story, builds a large boat and takes animals aboard to save them from a flood. The story in which Utnapishtim's deeds are told is the oldest epic poem in the world. Do you know what it is called?

⓮ In their research in Mesopotamia, have archaeologists discovered any evidence of a real, historical flood that could have given rise to the Noah/Utnapishtim stories?

⓯ The first archaeological evidence of human use of a written language comes from Mesopotamia. There scientists have found clay tablets covered with angular characters that were pressed into the clay while it was still soft. What do we call this kind of writing?

⓰ People who lived along the valley of the Nile in the land we now call Egypt invented another kind of writing. Ancient Egyptian is a picture language, using visual symbols to represent syllables and words. What is the technical name for the symbols used in this kind of writing?

17 For thousands of years, the way to interpret Egyptian hieroglyphics was forgotten, and archaeologists were unable to decipher the writings found on Egyptian tombs and temples. Then, in the early years of the nineteenth century, an important discovery was made by a soldier in Napoleon's army, which had conquered Egypt. He found a stone inscribed with two separate kinds of writing. One of the languages was Greek, the other the unknown language used in making hieroglyphics. As the inscriptions were studied, it was evident the texts were the same. The Greek translation thus furnished the key to the Egyptian characters. Here's a two-part question: What do we call the stone that unlocked the secrets of hieroglyphic writing? And who was the French linguist who deciphered it?

18 Most written languages today are alphabetic. That is, they use letters to represent the sounds of spoken words. Where and when was the idea of alphabetic writing invented?

19 In 1947 Bedouin tribesmen near the Dead Sea discovered ancient manuscripts that had been hidden long ago in a dry cave. These manuscripts, which may be more than 2,000 years old, contain previously unknown versions of prophetic writings in the Old Testament and are certainly one of the

most important archaeological finds of modern times. What are these manuscripts called?

20 One item of human manufacture has provided archaeologists more clues to the human past than any other. What is this commodity? Hint: It is used in the storage, cooking, and eating of food.

21 Wherever civilization developed on Earth, people invented money—which can be defined as anything that is used as a medium of exchange. Stones, seashells, and salt are among the commodities that have been used as money. Where was the use of metals as money first developed?

22 Coins of fixed value made of precious metal were first used by a fabulously rich king in Lydia (which is now part of the modern country of Turkey) about five-and-one-half centuries B.C. Can you name him?

23 What is the oldest surviving building constructed by human beings?

24 The great pyramids at Giza, in Egypt, have long fascinated travelers who stare in wonder at their imposing bulk. For what Egyptian ruler was the largest pyramid built, and how old is it?

25 The designers of the great Egyptian pyramids hoped to foil looters by installing elaborate security systems. False burial chambers and corridors led nowhere. Passages that did lead to burial chambers were blocked with huge blocks of stone. Maledictions written in hieroglyphics threatened disease and death to any who disturbed a Pharaoh's rest. Did any of the pyramid tombs remain unlooted?

26 The greatest archaeological treasure find of all time was made in Egypt in the 1920s. For the first time in recorded history, an unrobbed tomb (not a pyramid) was discovered and opened. Many rare and curious objects of costly material and workmanship were discovered. The most valuable of all were a solid gold coffin and a death mask of an ancient king, also made of solid gold. Though the ruler whose tomb had been discovered was a rather minor king, who had died at the age of eighteen, his name is very famous today. Do you know the king's name? And the name of the British archaeologist who made the find?

27 There is a city in modern-day Syria that is believed to be the oldest continuously inhabited city in the world. Can you name it?

28 The story of Heinrich Schliemann, a German archaeologist of the nineteenth century, is one of the most fascinating in the history of science. He became independently wealthy as a merchant trading in Russia. In 1868 he was able to retire from business and set out in quest of a long-cherished dream. He wanted to find a certain famous city of long ago. He used as his guide an epic poem written two-and-a-half millenia before his time. What was the poem? Who wrote it? And what was the famous city Schliemann found?

29 Modern archaeologists don't think much of the excavation methods used by Schliemann and other archaeologists of his time. Hiring gangs of local laborers, Schliemann directed a hasty excavation process that destroyed a lot of archaeological evidence. In the eyes of today's archaeologists, people like Schliemann were really "treasure hunters." Did Schliemann ever find any buried treasure?

30 In August of 79 A.D., an Italian volcano erupted in a great outpouring of ash, gas, and lava. The eruption buried two Roman cities so suddenly that people going about their daily llves died instantly. The buried cities became preserved in time. Today, many thousands of tourists visit these cities, parts of which have been carefully excavated by archaeologists. Can you name these two buried cities? Can you name the volcano that destroyed them?

31 Which ancient civilization was the first that is known to have such modern conveniences as indoor plumbing and flush toilets?

32 One of the most famous archaeological sites in the world is Stonehenge, on the Salisbury Plain of England. Theories about the people who built Stonehenge are many and varied. Some researchers maintain that the stones of Stonehenge were used as an astronomical calculator, to determine such dates as that of the summer solstice. How old are the oldest portions of Stonehenge?

33 Archaeologists have a scientific name for ancient stone structures such as Stonehenge. What do they call this type of architecture?

34 Archaeologists investigating an amazing burial site in China were astounded when they discovered a buried army of more than 6,000 soldiers drawn up in formation. Though these soldiers were buried more than 2,000 years ago, every detail of their physical appearance and the clothing they wore is perfectly preserved. How is this possible?

35 One of the most important concepts used by archaeologists to interpret their excavations is called the stratigraphic principle. What is this simple but scientifically useful idea?

36 What did the Ice Age have to do with how the first people got to what is now the Americas? How long ago did they get here?

37 The Incas of South America established an empire in what is now Colombia, Peru, Bolivia, and Argentina. They did not have wheeled vehicles, yet they built an impressive system of paved roads throughout their domain. What was the purpose of this road network?

38 Why didn't the Incas and Mayans use wheeled vehicles?

39 The planet Venus is sometimes seen as the "Morning Star," sometimes as the "Evening Star." How is this fact related to the conclusion that the ancient Mayans were great astronomers?

40 What kinds of animals, extinct by the time Europeans came to the Americas, were still around when the first Asiatics arrived?

41 Early Americans seem to have used only two kinds of animals as beasts of burden. What were they?

42 How many people were living in the Americas when Columbus landed in 1492: (a) 5,000,000, (b) 20,000,000, or (c) 50,000,000?

43 Can you name the famous American who made the first archaeological studies in the United States?

44 Archaeologists digging into a huge mound in western Illinois knew that the hill they were excavating had been made by people, not nature. But they weren't prepared for an astonishing discovery they made. What was it that so amazed them?

45 If you were traveling in the western United States and saw paintings and writ-

ings of early Americans on a rock face, would you know whether you were looking at pictographs or petroglyphs?

Answers

① *The oldest fossil bones considered to be those of creatures ancestral to humans have been discovered in Africa, and most scientists believe that Africa was the continent where humans evolved. A few researchers, however, argue for an Asian origin.*

② *She has been named Lucy, and is so called in scientific literature. Lucy was only about three-and-one-half feet in height. Scientists estimate that she was about twenty years old when she died.*

③ *Because their footprints were saved for our wondering view. Careful excavation by scientists has uncovered about fifty footprints made by two creatures, each walking on two legs. The footprints were made in a mud flat. The mud dried in the sun and was covered by ash from a nearby volcano, thus preserving the tracks.*

④ *It is called the hunter-gatherer stage of human prehistory.*

⑤ *It is estimated that the tools were covered by volcanic ash 1.89 million years ago. To put that in round numbers, which are easier to remember, we can say that humans have been making tools for almost 2 million years.*

⑥ *Remains of fires built by long ago Africans have been estimated to be 1,400,000 years old.*

⑦ *They are called baulks. The purpose of leaving them in place is to provide a cutaway view of the layers of deposits.*

⑧ *This ancient evidence of bronze working was found in Thailand.*

⑨ *Sites of human occupation dating about 300,000 years ago have been found in France and Spain.*

⑩ *Bone needles, with an eye for insertion of sinew thread, have been found at Cro-Magnon sites.*

⑪ *They were wheat and barley. These early "Turkish" farmers also planted peas and related vegetables such as vetch and lentils.*

⑫ *The Tigris and the Euphrates, which flow through modern day Iraq to the Persian Gulf.*

⑬ *"The Epic of Gilgamesh"*

⑭ *In 1930 while digging in the ancient Babylonian city of Ur, the British investigator, Sir Leonard Woolley, found a thick layer of clay he believed to have been deposited by a huge flood of the*

Euphrates River about 2900 B.C. He thought he had discovered evidence of the Biblical flood. Other scientists point out, however, that there is evidence of many floods in Mesopotamia, and that we cannot prove which one of them gave rise to the Biblical tale of Noah and his ark.

⑮ *Cuneiform*

⑯ *The symbols used by the ancient Egyptians are called heiroglyphics.*

⑰ *The stone is known as the Rosetta Stone, because it was discovered a few miles from an Egyptian village named Rosetta. The French linguist who cracked the code of the Rosetta Stone was Jean-Francois Champollion.*

⑱ *The idea of alphabetic writing was invented by Semitic people living in lands bordering the eastern Mediterranean about 1700 B.C.*

⑲ *They are known as the Dead Sea Scrolls.*

⑳ *The answer is pottery. The reason for the archaeological importance of pottery is that, unlike organic materials, ceramics last for many thousands of years.*

㉑ *In the Middle East. Metal money came into use there about 2000 B.C.*

㉒ *He was named Croesus. He issued coins of standard weight in both gold and silver. His name still survives in our language in the phrase "as rich as Croesus."*

㉓ *It is the 4,600-year-old step pyramid built as a tomb for an Egyptian king named Zoser.*

㉔ *The largest of the pyramids was built for Cheops (also known as Khufu) who ruled Egypt 4,500 years ago.*

㉕ *Researchers believe that all the pyramids were looted of their buried treasures long ago.*

㉖ *The king was Tutankhamen, commonly known as "King Tut." The archaeologist was Howard Carter.*

㉗ *The oldest continuously inhabited city is Damascus, capital of the modern Syria. It was first settled sometime before 2000 B.C.*

㉘ *The poem was "The Iliad." It was written by the Greek poet Homer. The city Schliemann found was Troy, the destruction of which is described in Homer's poem.*

㉙ *Yes, proving once again that the rich get richer, Schliemann actually found two treasures, the first during his excavations at Troy, the second when he excavated at Mycenae, on the mainland of Greece.*

(30) *The cities were Pompeii and Herculaneum. The volcano was, and is, Vesuvius. Unlike the cities, the volcano is still active. People attracted by the rich volcanic soil still farm its slopes.*

(31) *It was the Minoan civilization, ruins of which have been excavated on the island of Crete. The Minoans were prosperous traders who built lavish homes for their families.*

(32) *It is believed that the first stones placed at Stonehenge were brought to the site and put in position about 2800* B.C.

(33) *It is called "megalithic" architecture, from Greek word roots meaning "large" and "stone."*

(34) *Because the buried army consists of life-size ceramic figures. Each of the pottery warriors is an individual likeness, and each is painted with the proper insignia of military rank. The soldiers thus preserved were probably the palace guard of the Chinese Emperor Qin Shi Huang Di, who ruled China about 2000* B.C.

(35) *The stratigraphic principle is the basis for understanding archaeological sites where many layers of prehistorical evidence lie atop each other. The archaeologist logically assumes that the layers can be dated in sequence—that is, that the topmost layer is the most recent, and that each layer beneath is older than the one above.*

(36) *During the last Ice Age, so much of the water on Earth was locked up in glaciers and icecaps that the ocean level sank. A land bridge then connected Asia and North America across what is now the Bering Strait. Ancestors of all the Native American peoples crossed this land bridge. Archaeologists argue about how long ago this migration took place. Some say it happened about 20,000 years ago; others believe that "the first Americans" arrived 50,000 or more years ago.*

(37) *The Inca roads, which were carved out of mountainsides and crossed huge chasms on masonry bridges, were used to move military forces on foot and to enable relays of couriers to bear messages from one part of the kingdom to another. So swift were these couriers that it is said they could run 150 miles in a day.*

(38) *You know the answer to that about as well as archaeologists and anthropologists do. They can't explain why the Incas and Mayans didn't develop wheeled vehicles. It wasn't because they hadn't invented the wheel. They had. Toys with wheels have been found in Central and South American excavations.*

(39) *We know the Mayans had advanced astronomical knowledge because their records show that they plotted the movement of Venus with an error of only fourteen seconds a year.*

(40) *Mastodons, woolly mammoths, camels, and horses.*

(41) *Dogs were used as beasts of burden in North America, llamas in South America. The llama, which is a small animal, could manage to carry only about forty pounds. Dogs, which are even smaller, are known to have carried that much or more on their backs. Hitched to a travois, a small platform fastened to two poles, they could pull a lot more.*

(42) *Take your choice. Any one of the figures might be correct. Estimates by scientists vary from the lowest to the highest figure.*

(43) *He was Thomas Jefferson, the third president of the United States, who wrote a report on Indian ruins in Virginia.*

(44) *They unearthed the remains of what is believed to be the first city ever built in North America. This settlement, which we call Cahokia, was constructed by people known as the Mound Builders more than 1,000 years ago, before the time of the Mayan and Aztec civilizations. It is estimated that Cahokia had a population of 25,000.*

(45) *Memorize this little verse and you will always be able to identify them correctly: "Some elemental facts with which to be acquainted: A petroglyph is picked; a pictograph is painted."*

The Wonderful Human Body

The human body has been called the most wonderful machine known to science. A marvel of biochemical engineering, its self-repairing materials are miracles of molecular design. During its life span, the human body operates at amazing efficiency by taking in and utilizing a wide range of foods. With reasonable care, the average human body will last more than seventy years. In extraordinary cases, it has been known to keep on functioning for over 130 years.

What makes this human machine doubly remarkable is that it is an intelligent organism (or so we like to believe), being directed as it is by a fantastic biological computer, the brain, which awes and astounds scientists seeking to understand how it works.

Now, put your own biological computer on-line as you process these questions about the wonders of the human body.

1 How many miles of blood vessels does it take to keep all your body's tissues and organs supplied with oxygen-rich blood?

2 The year 1628 is important in medical history because of an experiment carried out by an English doctor, William Harvey. He tied a bandage very tightly around a man's arm. What great discovery did this lead to?

3 What's the largest organ of your body?

4 Is there a scientific explanation for the fact that many people like to sing in the shower?

5 What percent of your weight is made up by your bones—14 percent, 25 percent, or 40 percent?

6 What do you think? . . . Is your skull just one big bone?

7. Where are the smallest bones in your body located?

8. You've seen spectacular demonstrations in which a karate expert breaks a board or a brick with the side of a hand. You know that tremendous force must be involved. Why doesn't the karate expert's hand break?

9. Physiologists have determined that the average human fingernail grows about eight-tenths of an inch a year, faster in the summer than in the winter. In referring to the growth of fingernails, physiologists have spoken of "a nail second." What do they mean?

10. Perhaps you've heard the expression "water on the brain." What percentage of your brain is water?

11. Who was the famed artist/scientist credited with making the first accurate drawings of human anatomy?

12. Humans regulate their internal temperature by evaporative cooling through sweat glands, which are found all over the body. How many sweat glands does an adult human have?

13. We know that learning takes place primarily in the brain. Why then do we speak of learning something "by heart"?

14. In 1842 a daring American carried out the first surgical operation in which a

patient was anesthetized to eliminate pain. What was the anesthetic used in this procedure, and what operation was performed?

15 What organ in the body produces three quarts of acid juices in a day's time?

16 There are six to eight thousand taste buds in your mouth. How many basic tastes can they discern?

17 What is the hardest substance in your body?

18 If someone accused you of being lachrymose, what would it mean you were doing a lot of?

19 The ordinary American puts two or three teaspoons of salt into his or her system every day. How much salt is required in the human diet?

20 If you want to get extra iron in your diet, you should eat a lot of spinach, right?

21 How many bones are there in the human skeleton?

22 A Viennese doctor, Karl Landsteiner, won the 1930 Nobel Prize for medicine for a discovery that has saved countless lives. What was it?

23 Modern understanding of the mechanism of human pain says that nerves report damage of body tissues to the brain. It is the brain that "feels the pain." Is brain tissue itself subject to pain?

24 The average person spends something like eight out of every twenty-four hours sleeping. How much of that sleeping time does the average person spend dreaming?

25 All phone calls are transmitted by electrical impulses. The human brain also communicates information by electrical impulses. Do you think the daily number of electrical impulses generated by the world's phone system are greater than the number of impulses generated in a day by a single human brain?

26 When we inhale, we pump air into the lungs, where oxygen is extracted. Are there any muscles in the lungs?

27 Most foods are absorbed into the bloodstream through the walls of the small intestine. How long is the small intestine?

28 How many hairs are on the average human head? How about on the entire human body?

29 Do the fine hairs on your body serve any purpose?

30 At the age of two, a person has reached half the height he or she will be when fully grown. True or false?

31 Diabetes was once an incurable disease. About one person in fifty has it, but it can be controlled by the use of a drug called insulin, which was discovered in 1921 by two Canadian doctors, Frederick Banting and Charles Best. Can you name the bodily organ that produces insulin?

32 Do techniques such as yoga, meditation, prayer, biofeedback, imaging, and practicing deep relaxation have any proven medical benefit?

33 Are you right-handed or left-handed? Different estimates have been made of the percentage of people who are left-handed. It seems that about one person in seven may be naturally left-handed. Did you know that people usually also are "right-brained" or "left-brained"? Which side of the brain is usually dominant in right-handed people? Which in the left-handed?

34 Is butter more fattening than margarine?

35 There are about ten thousand nerve cells in the human body that control cycles of waking and sleeping. Where are they located?

36 How many different shades of gray hair have scientists distinguished on human heads?

37 Even before human infants begin to talk, scientists can tell what language

their parents speak just by listening to tape recordings of the babies' babblings. How?

38 The juices found in the human stomach are powerful chemical agents. They break food down into simpler chemicals that can be absorbed into the blood. Why don't these digestive juices dissolve the stomach itself? Or do they?

39 The human retina, located at the back of the eyeball, is an incredibly sensitive light detector. There are two kinds of cells in the retina that respond to light. They are called rods and cones. One of these kinds of cells can distinguish between different colors; the other sees only black and white. Is it the rods or cones that make color vision possible?

40 Did you ever wake up with a spider crawling on your face? It's creepy, isn't it? Now let's put you on the spot. In any given twenty-four-hour period, are the odds against or in favor of a member of the spider family being on your face?

41 What are the odds that two people could have identical fingerprints?

42 Can you get a suntan on a cloudy day?

43 Indianapolis race cars travel around the racetrack at a little over 200 miles per hour. Is this speed faster or slower than the transmission of electrical nerve impulses in the human body?

(44) A twentieth century Russian scientist demonstrated what is called a "conditioned reflex." He trained dogs to expect food when they heard the sound of a bell. Upon the ringing of the bell, the trained dogs would salivate whether or not food was present. Name this pioneer of modern physiology.

(45) What is the smallest living organism?

(46) In 1928 Alexander Fleming, a British bacteriologist, noticed something strange in his laboratory. A culture of staphylococcus bacteria had become contaminated with a blue-green mold. Dr. Fleming observed that around the growing mold, the staphylococcus organisms had died. Obviously the mold possessed some unknown antibiotic properties. Eleven years later, scientists would identify a substance secreted by the growing mold, and its manufacture would later save millions of human lives. What is this antibiotic that was first found in mold?

(47) In 1943 a Russian immigrant to the United States, Dr. Selman Waksman, discovered an antibiotic in a common mold that would be named streptomycin. Where does this life-saving mold grow in nature?

(48) A white fungus called *Phytophthona infestans* was responsible for great human misery in Ireland in the year 1845. It caused not

only widespread starvation but also a wave of Irish immigration to the United States. In what manner did the fungus change so many human lives?

49 *Aspergilus flavus* is a common mold that grows on many kinds of stored foods, such as peanuts, corn, wheat, and potatoes. What disease condition can this mold cause in people?

50 Present-day knowledge of chemical reactions would be impossible without scientific instruments that magnify the very small. The man who invented the microscope was a Dutch merchant who had no official scientific training. Do you know his name? Maybe. But—no peeking now—can you spell his name?

51 How much does your heart weigh?

52 How much does your brain weigh?

53 At what age does the human brain reach its full size?

54 We sometimes speak of the brain as "gray matter." Is the human brain really gray?

55 How many sounds can the human ear distinguish?

56 Children have a better sense of taste than adults. True or false?

57 The bump in the upper throat is called the "Adam's apple." What is the medical name of this structure?

58 There is a part of your body that, amazingly enough, does not get oxygen from blood but directly from the air. What is it?

59 How many cells does your body have? Can you guess within a few million?

60 The U.S. government is sponsoring the fifteen-year, $3-billion Genome Project. It has been called the most important medical research ever undertaken. What is the purpose of this project?

Answers

① *Your heart is constantly pumping blood through 100,000 miles of vessels.*

② *This led to the discovery that the heart is a pump. The tight bandage made a vein stand out. When Dr. Harvey touched the vein he could feel the blood pulsing through it. From this and following observations, Dr. Harvey was able to establish the way the heart circulates blood through the body's arteries and veins.*

③ *It's on the outside, not the inside. It's your skin, which, for an adult, averages eighteen square feet in area and weighs about six pounds.*

④ *Maybe some people sing in the shower because they think the noise of the falling water will mask sour notes. But here's a more scientific reason for shower-stall singing. Water falling through air creates negatively charged air molecules. These negative ions, scientists have discovered, make people cheerful, and cheerful people are more likely to sing.*

⑤ *Bones, on the average, account for only 14 percent of total body weight.*

⑥ *No. Your skull is actually made up of twenty-eight separate bones. Their wavy edges fit tightly together, somewhat like the pieces of a jigsaw puzzle.*

⑦ *The smallest bones in your body are in your ears. The bone called the* malleus *(hammer) is set moving by the eardrum. The malleus strikes against the* incus *(anvil). The incus is attached to the* stapes *(stirrup) that transmits vibrations to the inner ear, where mechanical energy is converted into nerve signals for processing by the brain. All three of these tiny bones are in the middle ear, which has a volume of only two cubic centimeters.*

⑧ *Because bones are amazingly strong. They bend under impact and distribute stress to muscles and other tissues. A bone, for instance, is forty times harder to break than concrete.*

⑨ *"A nail second" is the amount a fingernail grows in one second. How much is that? About .0000039 (39 ten-millionths) of an inch.*

⑩ *About 70 percent of your brain is water.*

⑪ *Leonardo da Vinci. He personally dissected cadavers to make sure his drawings were accurate.*

⑫ *Each adult human has around 2 or 3 million sweat glands.*

⑬ *Because it was once thought that the heart was the seat of human intelligence. This had been taught by Aristotle, the great Greek philosopher.*

⑭ *The anesthetic was ether, and the surgery was the removal of a tooth.*

⑮ *The stomach.*

⑯ *Just four—sweet, sour, bitter, and salty. Much of what we call taste is really smell.*

⑰ *The enamel on your teeth.*

⑱ *Crying.*

⑲ *Nutritionists figure a person needs only about a tenth of a teaspoon of salt a day. Excess consumption of salt can lead to many illnesses.*

⑳ *Actually, spinach contains no more iron than many other vegetables. And the body's iron absorption from spinach is inhibited by oxalic acid which is also found in this vegetable.*

㉑ *Either 206 or 207. (Some people have an "extra" vertebra.)*

㉒ *Landsteiner discovered the basic "blood types" that made transfusion of whole blood possible.*

㉓ *No. Brain tissue contains no pain receptors.*

㉔ *About two hours, or one-quarter of it.*

㉕ *It is believed that one human brain generates more electrical impulses in a single day than all the world's telephones put together.*

㉖ *No. Lungs have no muscles. The diaphragm and ribcage do most of the work of inhaling air.*

㉗ *Inside the body, the coiled small intestine is about eight feet long. If removed, however, it can easily be stretched to over twenty feet in length.*

㉘ *Scientists estimate that there are 125,000 hairs on the average human head—5,000,000 on the average human body.*

㉙ *Yes. When the hairs are moved—by air, touch, or pressure—they stimulate receptors that send appropriate signals to your brain.*

(30) *True.*

(31) *Insulin is produced in the pancreas, a large, many-branched organ that sits behind the stomach and discharges digestive substances into the small intestine.*

(32) *All the techniques mentioned have been scientifically demonstrated to improve the health of people using them.*

(33) *Since the right half of the brain controls the left side of the body and the left side of the brain controls the right side of the body, it is natural that right-handed people are usually left-brained and left-handed people are the only ones usually in their right minds.*

(34) *No. Margarine and butter contain just about the same amounts of calories and fat. Butter does, however, contain more cholesterol, which many nutritionists believe contributes to the development of heart disease.*

(35) *In the hypothalamus, which is a part of the brain. The hypothalamus controls many basic physiological functions.*

(36) *None. All gray hairs are really white. They look gray because they may be mixed with darker hairs.*

(37) *By analyzing the pitch and rhythm of the sounds the infants make. At about six*

months of age, babies are already successfully imitating the sounds of the language they hear, even though they speak no recognizable words.

(38) *Yes and no. The digestive process involves a balance between powerful acids and bases. Too much of one or the other leads to gastric injury and pain. Even a healthy person's stomach lining is renewed about every three days.*

(39) *The cones are the cells responsible for color vision.*

(40) *Greatly in favor. The vast majority of humans have miniature spiders, more accurately called mites, living in the sebaceous (oil-producing) glands of the facial skin. Don't worry, they're harmless.*

(41) *The FBI says that the odds are one in 64,000,000,000. So far the agency has collected more than 169,000,000 sets of fingerprints and has never found two to be identical.*

(42) *Yes. Eighty percent of the ultraviolet light that produces suntan penetrates clouds.*

(43) *The Indy race car wins, but not by much. Human nerve impulses have been timed at 180 miles per hour.*

(44) *Ivan Pavlov.*

(45) *A virus. The one that causes the common cold is so tiny that 3 quintillion of them weigh less than an ounce.*

(46) *Penicillin.*

(47) *This mold with antibiotic properties is* Streptomyces griseus. *It grows in dirt and piles of manure.*

(48) *Not by any disease condition of the human body. The fungus attacked the basis of the Irish food supply—the potato. The growth of the fungus was caused by successive seasons of unusually hot and wet summer weather.*

(49) Aspergilus flavus, *also called aflatoxin, is known to cause liver damage and even cancer in humans.*

(50) *The inventor of the microscope (in 1683) was Anton van Leeuwenhoek (or Leuwenhoek). His last name should be pronounced LAY-venn-hook.*

(51) *If you are an adult, your heart, which is approximately the size of your clenched fist, weighs about eleven ounces (310 grams).*

(52) *If it's average in size, it weighs about three pounds, or about 2 to 3 percent of your body weight. All that heavy thinking you do causes it to use 20 percent of all the oxygen carried in your blood.*

(53) *Your brain achieved 90 percent of its size by the time you were six years old. The 10 percent growth still to go was, however, pretty important. It occurs mainly by an increase in the number of connections between neurons. The more connections your brain has, the smarter it is. It is generally considered that the human brain achieves its full physical size at about sixteen years of age.*

(54) *In general, neural tissue (including the brain) is brownish-gray. Surprisingly, the two sides of the brain are different in color. The cells in the right side of the brain are much paler, more nearly white.*

(55) *Experiments indicate that an adult can distinguish 500,000 different sounds.*

(56) *True. You lose tastebuds as you get older.*

(57) *The Adam's apple is part of the human voicebox, or larynx. It is made of cartilage.*

(58) *The corneas of your eyes are the only parts of your body that have no blood supply.*

(59) *You probably didn't even come close. There are about 60 trillion cells in your body.*

(60) *Genome Project scientists are setting out to determine the exact arrangement*

of the 3 billion base pairs of the human DNA chain, which contains every individual's genetic heritage. If they succeed, doctors will someday be able to diagnose and treat many diseases that are incurable today.

Amazing Animals

We share the Earth with millions of other species of animal life. Nobody knows just how many species of animals there are, because the count is still going on.

Many scientific disciplines contribute to our understanding of animal life and behavior. People have been studying animals at least since the first Stone Age hunter made plans to answer the primordial human question, "What's for dinner?"

Let's test your zoological IQ.

1 Zoologists study animals, whereas botanists study plants. What is the difference between an animal and a plant?

2 Animal life depends on oxygen. We humans inhale ours in the air we breathe into our lungs. One class of creatures gets oxygen right through the skin. Can you name the skin-breathers?

3 The Swedish naturalist, Carolus Linnaeus, published a very important scientific work in 1758. In it he first propounded the modern method of giving scientific names to living creatures. In this system, each genus and species is given two names (usually Latin, or Latin in derivation). The human genus has such a scientific name—*Homo sapiens*. *Homo* means "man" or "human." What does *sapiens* mean?

4 The zoological definition of a species is a group of animals capable of mating with each other and producing offspring. In some instances, however, members of different species can breed. Often the results of such hybridizations are sterile and so cannot produce offspring themselves. One such animal is the mule. Which two species, bred together, produce mules?

5 Many millions of people have visited and enjoyed zoological laboratories, which can be found all over the world. You've probably visited one of these laboratories yourself. What was it called?

6 Where was the earliest known zoo located?

7 What was the first zoo built in the United States?

8 In their quest to understand animal life, zoologists study animals both captive and wild. Through examination of fossils some zoologists even study animals that are extinct. What is this branch of zoology called?

9 Which state of the U.S. has an extinct mammal as its state animal?

10 The fossil record shows that many creatures have come and gone upon our planet. The dinosaurs, for example, evolved about 225 million years ago and died out about 65 million years ago. But some creatures still living today are older than dinosaurs. There's a species of sea snail (scientific name: *Neopilina galatheae*) that's found off the shores of Costa Rica in the Pacific Ocean. Ancient fossils of this snail proved these animals to be the oldest known living species. How long has *Neopilina galatheae* lived on Earth?

11 You might have thought of this man as "just a philosopher." He lived in the fourth century B.C. His writings contain careful descriptions of the physical structure and life habits of animals. These writings qualify him as "the first zoologist." What was his name?

12 What is the fastest animal on Earth? Well, if you think for a moment, you'll realize that humans are the speed champions. But humans achieve their high velocities with mechanical contrivances. So let's talk about speed achieved solely with muscle power. As you might expect, the fastest travelers are birds. For sustained flight, the champion is a bird appropriately called the swift. For short distances, the record is held by the peregrine falcon while diving on its prey. How fast does the swift fly? How fast does the peregrine falcon dive?

13 Of all Earthbound creatures, the cheetah of Africa is the swiftest. In a short sprint after its prey, a cheetah has been known to attain a speed of 80 mph. But that's just a short burst of speed. What land animal can sustain the highest speed for long distances? (Hint: The animal locomotes on two legs.)

14 The continent with the greatest number of large animals that are potentially dangerous to humans is undoubtedly Africa. Which African predator is responsible for the most human fatalities?

15 Remember the "Big Bad Wolf"? He's a creation of the human imagination, of course. What about wolves? Do they deserve the fear they inspire in people? Not really. There is no record of any human having been killed by a wolf anywhere in the U.S. So, what do wolves really hunt?

16 Old-timers in the southwestern deserts of the U.S. maintained that some creatures, such as kangaroo rats, never drank water.

Scientists were skeptical. Surely, they thought, kangaroo rats must drink sometimes. Do they?

17 How many quills does a porcupine have?

18 Among other zoological characterizations of humans, we are vertebrates—that is, we have backbones. Zoologists divide vertebrates into seven classes: lampreys and hagfish, cartilaginous fish, bony fish, amphibians, reptiles, birds, and mammals. Which of these classes has the most species?

19 Of all the species on Earth, which is considered to be most closely related to humans?

20 Animal cousins of humans are the "anthropoid apes." There are four such species now living on Earth. How many can you name?

21 Humans are certainly more aggressive than any of their great ape relatives. Which human relative shows cooperative aggression most similar to our own?

22 Among large mammals, which mountain creature thrives at the highest altitudes?

23 Other than humans, which large mammal is believed to have the longest life span?

24 The lemming is a small rodent that engages in strange behavior. Vast numbers are often seen moving across the ground. They plunge fearlessly into rivers, lakes, and even the ocean, attempting to swim across. Many drown. How do scientists explain this suicidal behavior?

25 Do bears hibernate?

26 When European explorers reached what they would call the New World, none of the native North Americans had ever seen a horse. Where did the first horses originate?

27 Which mammal weighs as much as 200 pounds, has a sticky tongue up to a foot long, and digs with powerful claws in search of its favorite food—termites?

28 Most people are fascinated by birds. An ordinary person who pays attention to feathered creatures is called a bird-watcher; bird scientists are ornithologists. Ornithologists have roamed the world from pole to pole, cataloging bird species. How many have they counted so far?

29 What's the world's largest flightless bird? Your mind probably came up with that answer right away: "The ostrich, of course!" So here's something a little harder. When running at full speed across the countryside, how long a stride do you think an ostrich can take: (a) ten feet (b) fifteen feet, or (c) twenty-five feet?

30 Birds often crash into closed windows, injuring or killing themselves. Do they just fail to see the glass, or is there another reason for this suicidal behavior?

31 In the mid-nineteenth century, settlers in Utah were threatened with starvation when hordes of grasshoppers descended on their fields. The crops were saved by the coming of huge flocks of birds that ate the grasshoppers. What kind of bird saved the crops?

32 Members of the owl family hunt by night. Their large, sensitive eyes are known to provide excellent night vision. But scientific studies have shown that owls use a sense other than vision when hunting at night. How do they find their prey?

33 In the southern parts of Africa, farmers must constantly battle an airborne menace—a bird that flocks in huge numbers and descends on fields of crops and eats them bare. Breeding colonies of this bird can number in the tens of millions. What is this prolific bird?

34 Penguins are large flightless birds of the southern hemisphere. Once their ancestors lived in a subtropical environment, but, as the Earth changed, the penguin family had to adapt to more frigid and bleaker conditions. Their feathers are thick and oily, so they trap an insulating layer of air that keeps them from freezing in the cold waters where they swim. How many different species of penguins are alive on Earth today?

35 Some birds have sensitive nerves in their bills that help them find food. One such bird is a coastal bird of eastern North America. Its lower bill is quite a bit longer than its upper bill, and it flies low over the surface of the water. When the lower bill touches a fish, the bird's bill automatically snaps shut, thus catching dinner. What is this aerial fisherman?

36 Bird species have adapted to living in a wide variety of environments. Some, which hunt in water for their food, have evolved to swim underwater. The champion freshwater diver of the bird world has been accidentally captured in fishing nets at a depth of 180 feet, so we know it can dive at least that deep. It is a native of North America and once was much more common than it is today. It can still be found inhabiting the waters of northern regions. What is this diving bird?

37 Many birds migrate. Some travel long distances. Hummingbirds of North America, for instance, migrate to Mexico, the islands of the Caribbean, Central America, and even as far as South America. Their journeys may cover thousands of miles. But the longest migration of a bird species is truly global in scope. This bird travels all the way from the Arctic to the Antarctic and back each year, for a round-trip distance of more than 20,000 miles. Name this bird.

38 This is the world's smallest bird. The average male weighs only seven-hundredths of an ounce, so it would take more than 200 of them to weigh a pound. Can you identify this miniature aviator?

39 Hummingbirds dart swiftly through the air. They have the ability to hover in one spot to sip the nectar of blossoms. But can hummingbirds fly backwards?

40 The fastest recorded heartbeat of any creature is that belonging to the blue-throated hummingbird of the southwestern United States. How many times a minute does its heart pump?

41 Members of this bird species once made up 40 percent of the North American bird population. Breeding colonies were reported to cover forests over areas as large as forty square miles. Countless millions of them were shot, clubbed, and netted for food. Subjected to relentless hunting pressure, the numbers of this bird diminished steadily. It is believed that the last of its kind was shot in either 1899 or 1900, and what was once the world's most common bird became extinct. Name this vanished species.

42 When a peacock courts the lady of his choice, he displays for her delectation an array of feathered charms. How many colors are found in a peacock's tail feathers?

43 Approximately 150 species of poisonous snakes are considered to be dangerous to humans. Many people, knowing that some snakes are poisonous, are terribly afraid of all snakes. Yet there's a creature whose venom kills many more people each year than the number who die from snakebite. What is this common killer? (Hint: It has five eyes and is capable of X-ray vision)

44 What reptiles lives only in the United States and on branches of the Yangtse River in China?

45 There is an insect that may be the toughest creature on Earth. Entomologists who have studied this bug report that it can withstand 100 times the radiation that would kill a human. Cut off the head of this insect and it will stay alive for two or three weeks. What is this super bug?

46 You may have noticed that mosquitoes don't come zooming in on a cool day the way they do on a warm one. This is because of the fact that mosquitoes are cold-blooded creatures—that is, their body temperature changes to match that of the air around them. When they get too cool, mosquitoes can't keep their wings beating. What is the temperature below which mosquitoes are unable to keep flying?

47 Do male mosquitoes bite human beings?

48 How many legs does a centipede have?

49 How many individual ants exist on the surface of the Earth at a given moment? Obviously, nobody has ever counted them one by one, but a scientific estimate has been made. So, how many ants: (a) one billion, (b) one trillion, or (c) one quadrillion?

50 For many people, worms are an unpleasant subject. There are three classes of creatures called worms—flatworms, roundworms, and segmented worms. Roundworms are among the most numerous complex creatures on Earth. Most are too small to be seen by the naked eye and live beneath the surface of the soil. If you picked up a handful of garden soil, how many roundworms, on the average, could you expect to find inhabiting it?

Answers

1 *The basic difference between an animal and a plant is that (in almost all cases) animals are capable of moving themselves around and plants are not.*

② *Amphibians are the creatures that get the greatest percentage of their needed oxygen through the skin. Some frogs absorb as much as 70 percent of their oxygen that way.*

③ *Looking at history or current affairs, you might not think it very accurate, but* sapiens *means something like "wise."*

④ *A mule is the offspring of a mating between a female horse and a male ass.*

⑤ *It was probably called a zoo (which is an abbreviation for zoological garden). One of the purposes of zoos is the scientific study of their animal residents.*

⑥ *In Egypt. It was built about 1500 B.C. by Queen Hatshepsut.*

⑦ *New York City's Central Park Zoo, which opened in 1864.*

⑧ *Paleontology.*

⑨ *California's state animal is the California grizzly, a creature that no longer exists.*

⑩ *Fossils of* Neopilina galatheae *are believed to be about 50 million years old.*

⑪ *Aristotle.*

⑫ *The swift achieves sustained flight of approximately 100 mph. The peregrine falcon in a dive has been clocked at a blistering 180 mph.*

(13) Did you guess that humans are the marathon champs? You're right and wrong. Because they can carry food supplies with them (and so don't have to stop to eat), humans are the champions at really long distances. But in a single run, the distance champion may be the kangaroo. When hopping at full speed, kangaroos exceed 20 mph, a rate of travel they can maintain for hour after hour.

(14) The deadliest African animal is not the lion, nor is it any land animal. It is a river-dwelling reptile. Crocodiles kill more humans than any other African predator.

(15) Scientific study of wolves shows that the main article of their diet is rodents—small creatures such as rabbits and mice.

(16) No. Kangaroo rats get all their needed moisture from seeds they eat.

(17) Various zoologists have actually counted them and found that the average porcupine has about 25,000 quills.

(18) There are more species of bony fish than there are of other kinds of vertebrates. There are approximately 13,000 species of bony fish swimming the waters of our world. By comparison, there are only about 4,000 species of mammals.

(19) The chimpanzee. Geneticists have determined that 98 percent of the genetic material of chimpanzees and humans is identical.

⑳ *The four species of anthropoid apes are the orangutan and the gibbon, which live in Asia, and the chimpanzee and the gorilla, which live in Africa.*

㉑ *Baboons run in troops as large as 300 individuals. If a troop is threatened, males attack the source of danger while females and young retreat.*

㉒ *The Tibetan yak. Its thick hair enables it to live as high as 20,000 feet above sea level.*

㉓ *The elephant. A life span of sixty-five years or more is common among pachyderms.*

㉔ *It is believed that population pressure triggers a migratory instinct. The lemmings are seeking a new place to live.*

㉕ *No. Some bears sleep deeply during cold months, but do not hibernate. Some true hibernators are bats, prairie dogs, dormice, and hedgehogs.*

㉖ *Horses originated in North America millions of years ago. But 10,000 years ago, when the Amer-Indians arrived in the western hemisphere, this animal had already been extinct for quite some time.*

㉗ *The aardvark, native to southern Africa.*

(28) *Ornithologists count the number of known bird species at about 8,580.*

(29) *The correct answer is (c). Each step of an ostrich at full run covers about twenty-five feet.*

(30) *It is believed that birds crashing into windows are trying to protect their territory. A bird sees its own reflection, mistakes it for another bird moving in on its turf, and dashes at its own image. Talk about being mad at yourself!*

(31) *Strangely, though Utah is far from the sea, the rescuers were California gulls, no doubt attracted by the plentiful food supply.*

(32) *Owls hunt by hearing. Their ears are sensitive enough to guide them to mice and other small rodents that make up most of their diet.*

(33) *It is the quelea (pronounced KWEE-lee-uh).*

(34) *There are fifteen different species of penguins alive today.*

(35) *It is a large bird of the tern family, the black skimmer.*

(36) *It is the common loon. Incidentally, the phrase "crazy as a loon" refers to this bird, because of its wild cry, which sounds somewhat like insane human laughter.*

(37) *It is the Arctic tern.*

(38) *The world's smallest bird is the Cuba hummingbird.*

(39) *Yes.*

(40) *The blue-throated hummingbird's heart beats 1,200 times a minute when it is active.*

(41) *The bird species is (or was) the passenger pigeon.*

(42) *If you named blue, green, black, or yellow, you probably think the display feathers of a peacock are tail feathers. Not so. The true tail feathers are underneath the colorful ones and are short feathers of drab coloration.*

(43) *Bee stings kill many more people than snakebites do. With their five eyes, bees perceive X rays, ultraviolet light, and polarized light—all invisible to humans.*

(44) *The alligator. For extra credit, answer this: How do you tell an alligator from a crocodile? Very carefully, of course.*

(45) *The common cockroach.*

(46) *Mosquitoes cannot operate when the air temperature is less than sixty degrees (Fahrenheit).*

(47) *No. Only female mosquitoes drink blood. Males feed entirely upon the juices of plants.*

(48) *Fooled you. No known centipede has 100 legs. The "leggiest" has ninety-five pairs of legs; the one with the fewest legs has only thirteen pairs.*

(49) *The correct answer is (c). The biggest number is right this time. Scientists estimate the number of ants on Earth at one quadrillion (1,000,000,000,000,000.)*

(50) *Don't be horrified, but helminthologists (that's what scientists who specialize in worms are called) say that if you searched carefully with a microscope, you'd find about 1,000 roundworms in that handful of dirt.*

Green Wonders

Without plants, animal life, including that of humans, would not be possible on Earth. When scientists look back more than 2 billion years, they see a time when our planet's atmosphere contained no oxygen. Primitive plants produced the first oxygen and so set the stage for the development of animal life.

And still today, plant life is essential to the very survival of ourselves and all other animals. Plants produce not only the oxygen we must breathe, but they also provide, directly or indirectly, all the food we eat. Now let's find out what you know about plants, the green wonders that make life possible.

❶ Once upon a time, about 280 million years ago, all the continents on Earth were gathered into a supercontinent that geologists called Pangaea. What kinds of flowers bloomed on Pangaea?

❷ How long ago did the first flower bloom on Earth?

❸ The first plants (like the first animals) evolved and lived in the ocean. How long ago did the first land-dwelling plants develop?

❹ The period of human history we call the Industrial Revolution was made possible by plants that died hundreds of millions of years before, for it was coal that fueled the factories. And coal, as you know, consists of the remains of plants that grew and died in swamps and lowlands of long ago. What kinds of plants formed the fossil fuel deposits that made modern civilization possible?

❺ The first seed-bearing plants developed about 225 million years ago, at the start of a geological era called the Mesozoic. Descendants of these plants are still alive today. What kind of modern plants are related to the first seed bearers?

❻ The first conifers and the first dinosaurs developed at about the same time. Some paleontologists believe that the dinosaurs were responsible for the development of such large conifers as sequoias and redwoods. How so?

7 What is the largest plant on Earth?

8 Are sequoias the tallest trees on Earth?

9 Sequoia trees were named after a famous American. Was he (a) a famous botanist who was the first to explore California, (b) a military man who subdued native peoples of the Pacific Northwest, or (c) an Indian chief?

10 Most plants need sunlight to grow. They convert solar energy into vegetable mass by an amazing chemical process. What is this process called?

11 You probably know that a chemical called chlorophyll is essential to most plants' conversion of sunlight into chemical energy. You probably also know that chlorophyll is what gives foliage its green coloration. Knowing these facts, can you figure out whether chlorophyll uses green wavelengths of light to produce food?

12 Many plants exhibit a behavior known as phototropism—that is, they grow toward a source of light. Researchers have discovered that plants respond to only one color of the spectrum in this light-seeking behavior. What is the color?

13 All living creatures consist of one or more structures called cells. Who was the first scientist to describe cells?

14 How can you determine the number of cells in a particular deciduous tree?

15 Why do broad-leaved trees (such as oaks, maples, elms, and so on) shed their leaves in the fall, while narrow-leaved trees (such as pines and other evergreens) keep their leaves year-round?

16 Compared to humans, many trees have very long lifetimes. Some redwoods and sequoias are believed to be more than 3,000 years old. Specimens of the bristlecone pine (another venerable inhabitant of the western U.S.) are nearly 5,000 years old. Does that make them Earth's oldest living things?

17 Seeds have amazing properties. Seeds of desert plants, for instance, can lie dormant in dry desert soil for many years, sprouting only when unusually heavy rains provide enough moisture for growth. Arctic plants have also adapted to bleak conditions. The oldest seeds ever known to sprout were those of a flower called the Arctic Lupin. How many years had these seeds lain frozen before they were planted by scientists?

18 What plant has the largest seeds?

19 What is the difference between a fruit and a vegetable?

(20) What percentage of the calories consumed by humans is derived from plant sources? What percentage of protein?

(21) How many species of plants are agriculturally cultivated in the world today?

(22) What is the science of agriculture called?

(23) What important agricultural plant was first grown high on the cold mountainsides in South America, at an altitude of more than 11,000 feet above sea level?

(24) What's the most caloric fruit eaten by humans?

(25) Many people find edible mushrooms to be extremely tasty. How can you tell if a fungus you find is edible or poisonous?

(26) What's the difference between a mushroom and a toadstool?

(27) A researcher once measured a certain part of a certain edible plant and found it to be 387 miles long. What part of the plant did he measure?

(28) In what way are weeds different from other plants?

(29) Why is it wrong to call a banana tree a banana tree?

30 A farmer might be glad to see a thunderstorm coming because of the rain that will water his crops. But why should he also welcome the lightning that will accompany the storm?

31 Many drugs used in medicine are derived from plants. One of the most common pain-killing drugs is aspirin, or, more technically, acetylsalicylic acid. Aspirin tablets became available in the 1890s, but people had been using "herbal aspirin" for thousands of years before then. What common plant contains the pain-relieving chemical found in aspirin?

32 Should you be worried about the Kudzu menace?

33 You know that bees and other insects play an important part in fertilizing plants by carrying pollen from one plant to another. There is a bird that does the same job. Can you name it?

34 Does the world's smallest flowering plant grow on land or in the water?

35 How long are the longest leaves of the plant kingdom?

36 Seeking water and nutrients, plant roots penetrate far into the soil. How deep were the deepest roots ever studied by science?

37 Many kinds of grasses are very important in Earth's ecology. Think of the

importance to humans of grains such as wheat, rice, corn, oats, barley, and rye (which are all grasses). The world's tallest grass can reach a height of 130 feet. Do you know what it is?

38 The world's largest flower grows on the Rafflesia plant of Indonesia. Is this bloom (at its largest) one, two, or three feet in diameter?

39 How long can a cactus go without water?

40 What do you think?... Do aquatic plants have roots?

41 How many species of plants are alive on Earth at the present time?

42 You know about pitcher plants and Venus's-flytraps, which capture insects and extract nutrients from their bodies. How many species of carnivorous plants have been identified?

43 Farmers and gardeners of today often find they must apply commercial fertilizers to the soil in which their plants grow. When you buy a fertilizer, it is described by a series of three numbers, such as 4-12-4, or 20-40-10, and so on. What do these numbers mean?

44 Soils that have become depleted in fertility can sometimes be restored to health by growing plants called legumes. How do legumes increase soil fertility?

45 An eighteenth-century American was called the greatest botanist in the world during his lifetime. Who was he?

46 Name the famous Swedish botanist who developed the modern scientific nomenclature to describe different species of plants.

47 His scientific work was ignored in his lifetime, but today the whole world knows the name of Gregor Mendel, whose experiments with the heredity of pea plants founded the modern science of genetics. What did Mendel do for a living?

Answers

1 *There were no flowers on Pangaea, because flowering plants had not yet evolved.*

2 *Nobody knows for sure, but the oldest fossils of flowering plants date from about 135 million years ago.*

3 *The first plants to live on land made their homes on ocean shores about 400 million years ago.*

4 *The great coal deposits of Europe and America were formed primarily from the fallen remains of mosses and ferns—many of which were the size of today's trees.*

5 *They are the conifers, or cone-bearing plants, such as pines, firs, and spruces.*

⑥ *Scientists believe that the giant conifers grew tall to escape the munching of giant dinosaurs.*

⑦ *The world's largest plant is a giant sequoia tree growing in northern California. The tree, which is called "the General Sherman tree," is 273 feet tall. The thought would horrify anyone with an ecological conscience, but it has been calculated that "General Sherman" contains enough wood to make 5 billion matches.*

⑧ *The tallest sequoia ever measured was named "Father of the Forest." It was over 400 feet tall when it was cut down by loggers. The tallest living sequoia is almost 370 feet tall. There are rumors of eucalyptus trees in Australia that equal or exceed this height, but so far no such eucalyptus has been measured by scientists.*

⑨ *The correct answer is (c). The tree is named after the Cherokee Indian leader Sequo-ya, who was the first person to develop an alphabet for an Amer-Indian language.*

⑩ *It is called photosynthesis. The roots of that word mean "put together with light."*

⑪ *No, it does not. Chlorophyll cannot absorb green wavelengths of light. Instead it reflects green light, and that is what we see when we look at the leaves of most plants.*

⑫ *Only blue wavelengths of light cause plants to change the positions of their stems and leaves.*

⑬ *He was the English scientist, Robert Hooke, who in 1665 examined thin slices of cork under a microscope and described the cell structure he saw. Hooke was also the first scientist to use the word cell.*

⑭ *The calculation might take you a while. Scientists estimate that the average deciduous leaf contains about 40,000,000 cells. To find the number of cells contained in the leaves of the tree, therefore, you'd have to count the leaves and multiply by 40 million. Now you also have to figure out how many cells are in the roots and trunk of the tree. A rule of thumb is that the woody parts of a deciduous tree contain about twenty times as many cells as the leaves do.*

⑮ *The primary reason is to conserve water. Broad leaves transpire (evaporate) much more water into the air than do narrow leaves. As winter comes, a tree's water supply is cut off by freezing of the soil. Therefore, broad-leaved trees must shed their leaves or they would suffer death from dehydration.*

⑯ *No. The oldest living creatures are not trees at all. A lichen found in Antarctica is believed to be at least 10,000 years old and may have been alive much longer than that. At the other "end of the Earth," some lichens in Alaska are said to be at least 9,000 years old.*

⑰ *Scientists believe that seeds found in frozen soil in the Canadian Yukon were somewhere between 10,000 and 15,000 years old when they were dug up. Planted under more favorable conditions they successfully sprouted.*

⑱ *The world's largest seeds are found on a kind of palm that grows in the Seychelles Islands of the Indian Ocean. It takes seven years for this coconut-like fruit to mature to its full-grown weight of more than forty pounds.*

⑲ *A fruit is (botanically speaking) the ripened ovary of a seed plant and its contents. Many of the edible plant products we call vegetables are, therefore, really fruits. Generally we call sweet, pulpy, seed-containing foods fruits. We reserve the word* vegetables *for the leaves, stalks, and roots that make up much of our diet and for the not-so-sweet foods, such as cucumbers, which are really fruits.*

⑳ *It has been calculated that 88 percent of the calories and 80 percent of the proteins consumed by humans are obtained from plants.*

㉑ *About 200 species of plants are presently cultivated by humans. Most of our food supply, however, is derived from a mere fifteen species.*

㉒ *It is called agronomy.*

㉓ *The potato. It was first grown in the Bolivian Andes.*

(24) It's the avocado, which provides more than 740 calories per pound. So if you're trying to lose weight, avoid avocados. Try cucumbers instead. This least caloric of fruits contains only about seventy calories per pound.

(25) It is sometimes difficult for even mycologists (that's what mushroom experts are called) to distinguish between edible and poisonous varieties. They recommend that nonexperts eat only those mushrooms sold in stores.

(26) There really isn't any. Toadstool is a popular term for any kind of poisonous or supposedly poisonous mushroom. It is not a term used by scientists.

(27) He measured the root system of a single rye plant. He calculated that this one plant had more than 13 billion individual roots!

(28) Weeds are different only in the eyes of humans. A weed is defined simply as "a plant humans consider undesirable."

(29) Because, by botanical definition, trees have woody stems. Banana plants have no such woody material, even though they grow as large as many trees.

(30) Lightning plays an important role in creating nitrogen compounds that contribute to plant growth. The estimated 15,000,000 thun-

derstorms that occur on Earth each year create 100,000,000 tons of nitrogen fertilizer.

(31) *The willow tree. The writings of the earliest physicians recommend a tea made of willow bark to relieve pain and reduce fever. Incidentally, doctors have no really good explanation of just how willow tea or aspirin tablets work in the human body.*

(32) *If you live in the southeastern section of the United States, maybe you should be. Kudzu is a fast-growing vine that was imported from Japan to the U.S. in 1876. It was just an ornamental plant until scientists discovered that it could be fed to livestock and that it added nitrogen to the soil. The U.S. Department of Agriculture convinced farmers in the Southeast to plant Kudzu. It went wild, blanketing farmlands, choking out shade trees, and covering utility poles and buildings. Every year farmers, corporations, and government agencies spend millions of dollars trying to get rid of the troublesome Kudzu.*

(33) *Hummingbirds, which live on nectar produced by flowers. To supply its daily energy needs, it has been calculated that a hummingbird must visit more than 1,000 flowers a day. The fact that in flitting from flower to flower they carry with them the fertilizing pollen makes them important contributors to the needs of the plant world.*

(34) *In the water. It is a floating duckweed (scientific name:* Wolffia arrhiza*). The*

whole plant averages less than a millimeter (one-twenty-fifth of an inch) in diameter.

(35) *Leaves of a tree called the Raffia Palm can grow to be more than sixty feet in length.*

(36) *Roots of a wild fig tree in Africa were found to grow down to a depth of almost 400 feet below the surface of the ground.*

(37) *It is a species of giant bamboo. It has been observed to grow an astounding three feet in a single day.*

(38) *The giant bloom attains a diameter of three feet.*

(39) *It varies from one kind to another, but many desert cactuses can hold a two-year supply of water in their tissues. One California specimen was uprooted and left lying on a bench in a laboratory for six years before it died.*

(40) *Most species of aquatic plants do have roots, even though they don't need them to obtain water. The roots of aquatic plants serve to hold them in place.*

(41) *Botanical scientists put the number of plant species on Earth at about 350,000.*

(42) *Worldwide, about 500 species of carnivorous plants have been catalogued.*

(43) *The three numbers refer to the three elements necessary for plant development and growth. The first number refers to the percentage of nitrogen in the fertilizer, the second to the percentage of phosphorus, the third to the percentage of potassium (sometimes called potash).*

(44) *Legumes increase soil fertility because they can combine nitrogen in the atmosphere (where it is of no use to most plants) with hydrogen to form amino acids, the building blocks of plant proteins. Legumes accomplish this fertilizing feat with the help of a certain bacteria, called* Rhizobium, *that live in their roots.*

(45) *The name of this famous American botanist was John Bartram, who lived from 1699-1777. Bartram traveled extensively in the southeastern U.S., collecting specimens of many plants that had been previously unknown.*

(46) *He was born Carl von Linne, but is more often referred to by the Latinized version of his name, Carolus Linnaeus. This is fitting, since it was Linnaeus who began the practice of giving Latinized names to species.*

(47) *Gregor Mendel was an Augustinian monk.*

Number Knowledge

The beginnings of mathematics most likely go back to long before recorded history. The most primitive people ever studied by anthropologists all have had some kind of counting or numbering system. The simplest mathematical systems known are still used by certain tribes who have only three words for expressing quantities—words meaning one, two, and many.

We, being civilized, and therefore living complicated lives, have need of a lot more "number words" and "number ideas" than that.

How do you rate your numerical know-how? Don't worry; none of the following questions involve any com-

plex calculations. They will just tell your basic knowledge of mathematical history and terminology.

1 You undoubtedly know that a prime number is a number that is divisible only by itself and 1. What is a Mersenne number?

2 And what is a perfect number?

3 How many perfect numbers are there?

4 How many prime numbers have been found?

5 Once upon a time, mathematicians worked laboriously, sometimes for years, with pencil and paper, to find things like prime, perfect, and Mersenne numbers. Now, of course, they use computers. The newest prime number was found by a Cray-2 supercomputer after nineteen hours of searching. As for what this knowledge might do for science, one mathematician jokingly put it this way: "Finding the largest known prime number is an enormously prestigious scientific feat of absolutely no practical significance." Why, then, do mathematicians engage in such searches?

6 A *googol* is a mathematical term for a really large number. Just how big is a googol?

7 When is a billion more than a billion?

8 Suppose you had a billion U.S. dollars, then suppose you decided to live in luxury and spend $1000 every day. Would you live long enough to spend the billion?

9 A billion years is a long time. Suppose you could push a button once every second, eight hours a day. Could you live long enough to push the button a billion times?

10 A famous mathematician made a practice of walking on the beach and using a stick to write equations on the sand. He was doing this one day when he was attacked and killed. Who was he?

11 A noted English mathematician became far more famous for a fantasy novel he wrote than he did for his mathematical studies. Who was he and what was the book that he wrote?

12 Everybody knows the mathematical expression $e = mc^2$. But do you know what it really means?

13 Perhaps you know that a thousandth of a second is called a millisecond and that a millionth of a second is called a microsecond. *Milli* and

micro are prefixes that mean "one one-thousandth of" and "one one-millionth of." Do you know the prefixes that enable you to make words that mean one billionth of a second? One trillionth of a second? One quadrillionth of a second? One quintillionth of a second?

14 Let's find out what you know about the history of mathematics. What's the oldest evidence we have for the use of a mathematical symbol for zero?

15 What land is the birthplace of our basic decimal system of writing numbers?

16 Do you know what *pi* () is in mathematics? The Greek letter is used to express the relationship between the radius of a circle and its circumference. Archimedes worked out that *pi* was somewhere between $3\frac{10}{71}$ and $3\frac{1}{7}$. A more precise value for *pi* is one you may have learned in school—3.1416. Which mathematician figured out this value of *pi,* which is still used today?

17 In the mid-fifteenth century, a German mathematician was the first to develop the concept of negative numbers. He also popularized the plus and minus signs (+ and −) that we use today. Yet few people know his name. Do you?

18 In the seventeenth century, two of the greatest mathematicians who ever lived waged a long and bitter controversy, each claiming to be the inventor of the kind of mathematics we call calculus. Who were these two mathematical giants?

19 The programs that run today's computers are constructed with the aid of a kind of algebra that bears the name of the English thinker who invented it. Who was the thinker, and what is the name of the algebra he devised?

20 In 1883 George Babbage, an English mathematician, designed a device that he called an "analytical machine." Babbage's invention was the forerunner of one of today's most important technologies. What would we call Babbage's invention today?

21 A famous Roman emperor authorized the adoption of a new calendar for the year. It divided the year into twelve months and provided for leap years, which had an added day. The calendar was used for more than sixteen centuries. Who was the emperor, what do we call his calendar, and what month in the year was named after him?

22 Which mathematician first calculated that it might be possible for humans to put an artificial satellite into orbit around the Earth?

23 We call our numbers Arabic numerals. Is that because they were invented by Arabs?

24 Many famous mathematical ideas have been formulated by studying gambling and games of chance. The basic mathematical principles of these subjects were figured out by the Italian mathematician Geronimo Cardono in the sixteenth century. How well do you understand gambling odds? Suppose we select six people at random. What are the chances of any two of them having been born in the same month? Six people, twelve months. Maybe it's just an even bet?

25 Bookkeeping is a branch of mathematics used by businesses and governments all over the world. The modern method of keeping books is called double entry bookkeeping. Who invented it?

26 In 1846 one of our solar system's planets was discovered with the aid of mathematics. Which planet was it, and how did mathematics tell astronomers that it existed, and where?

27 Someday the world chess champion will be a computer. At least that is the belief of both chess players and computer experts. Computers are formidable chess players because they can quickly analyze many thousands of possible move sequences. And there's a lot to analyze. How many different ways do you think there are for two chess players to make the first four moves (four on the part of each player) of a game?

28 A deck of playing cards contains fifty-two cards. Suppose we set out to figure how many different ways those fifty-two cards could be arranged in sequence. Are there more or less than a billion possible arrangements?

29 The so-called Arabic numerals are much easier to work with than the Roman numerals they replaced. But when they were introduced to Europe in the 1300s, Arabic numerals were illegal. Why was their use forbidden?

30 Under what conditions does ten plus ten equal 100?

31 Computers and electronic calculators have become so common in our world that students have forgotten what it's like to carry out tedious mathematical calculations with pencil and paper. The dream of building machines to take away the drudgery of this kind of work is not a new one. Do you know what the oldest successful calculating machine is?

32 Let's say you set up a contest between a skilled user of the abacus and a computer whiz using the fastest modern computer. You ask them both to add a long column of numbers to see who comes up with the answer first. Who do you think would win the race?

33 With the advent of the pocket calculator, the slide rule has become a museum curiosity. But during its 350-year heyday, this ingenious but simple calculating device enabled its users to quickly solve many kinds of mathematical problems. Who invented the slide rule?

34 Probably no other mathematician has had as great an influence as the great Greek geometrician, Euclid. His textbook on geometry is still used today, more than 2,000 years after he wrote it. In his own day, Euclid was a famous teacher. Where did he live and work?

35 This French mathematician was a true child prodigy. His father was a tax assessor who, to help his son with his calculations, designed the first working mechanical adding machine. By the time the boy was eleven years old, he had worked out his own proofs to many of Euclid's geometrical propositions. Do you know his name?

36 There were many attempts made in the early part of the twentieth century to design and build calculating and computing machines. It is generally considered that ENIAC (the letters stand for Electronic Numerical Integrator and Calculator) was the first successful electronic computer. It was built by Dr. John Mauchly and others at the Moore School of Engineering in Pennsylvania. In what year was ENIAC's awesome computing power first switched on, thus ushering in the computer age?

37 Like all early computers, ENIAC was huge. It incorporated more than 18,000 radio tubes and used large amounts of electricity. This made such computers very expensive. But there was an invention on the horizon that would make it possible to greatly shrink the size of, and power requirements for, computers (and other electronic equipment). What electronic device, made from silicon, would revolutionize computers?

38 Many people have asked and wondered whether a computer will ever be able to "think." Like many questions, the answer to this one depends on a definition. What is "thinking?" Human thinking seems to go beyond mathematics and logic and incorporate an intuitive component. All computers built or dreamed of until now are purely logical. So, as we leave this world of mathematics and computation, let's pose another question. A famous cyberneticist (a person who studies artificial intelligence) has proposed a challenge: Will it ever be possible to build a computer whose communications cannot be distinguished from those of a human? If a machine ever meets this challenge it will be said to have passed a test that bears the name of the computer scientist who thought it up. What is the name of this test?

Answers

(1) *A Mersenne number is a prime number that is expressed in the form of 2 multiplied by itself a certain number of times, minus 1.*

Mersenne numbers are named after a seventeenth century monk who spent his cloistered time hunting for prime numbers.

② *A perfect number is one that is the sum of every one of its factors, excluding itself. Take the number 28 as an example. It is divisible by 1, 2, 4, 7, and 14. Add those together and what do you get? Yes—28.*

③ *Recently, the thirty-second one was found. The number has 455,663 digits!*

④ *In 1992 a group of British mathematicians discovered the largest prime number known as of that time. It has 227,832 digits.*

⑤ *For the sheer joy of doing it, and of knowing.*

⑥ *A* googol *is the term for the number 1 followed by 100 zeroes, or written another way,* 10^{100}.

⑦ *Peculiar as it may seem, there are different definitions of a billion. An international billion is 100 times more than a U.S. billion.*

⑧ *You'd be getting pretty old. At $1000 a day, it would take more than 27,000 years to spend $1 billion.*

⑨ *Yes. You would have to be awfully dedicated, but if you pushed a button once every second, eight hours a day, no weekends or holidays off, it would take more than ninety-five years to push that button a billion times. Want to try for a trillion?*

⑩ *The mathematician who met his death while writing equations in the sand was Archimedes. In addition to being a mathematician, Archimedes, a citizen of Greece, was also an engineer. While in the employ of the king of Syracuse, on the island of Sicily, Archimedes designed and supervised the construction of numerous devices of war. When Roman soldiers conquered the city, Archimedes was slain, although command had been given that he be spared.*

⑪ *The mathematician was Lewis Carroll, who wrote "Through the Looking-Glass, and What Alice Found There," commonly referred to as "Alice in Wonderland."*

⑫ *In the early part of the twentieth century, Albert Einstein realized that mass could be converted into energy. The formula* $e = mc^2$ *means that if a given amount of mass is converted into energy, the amount of energy obtained can be calculated thusly:* e *(the energy) will be equal to* m *(the mass converted) times* c *(the speed at which light travels per second) squared.*

⑬ *A nanosecond is a billionth of a second. A picosecond is a trillionth of a second. A femtosecond is a quadrillionth of a second. An attosecond is a quintillionth of a second.*

⑭ *Archaeologists have found clay tablets on which Babylonian scribes used a symbol for zero. They were made in the fourth century* B.C.

⑮ *The first record we have of a decimal number system came from northern India about 1500 years ago.*

⑯ *His name was Ptolemy. He lived in Alexandria, Egypt, while it was under the rule of the Roman Empire.*

⑰ *He was Michael Stifel.*

⑱ *They were the Englishman, Isaac Newton, and the German, Gottfried Leibniz. Newton did not publish his discovery until almost twenty years after he made it. In the meantime, Liebniz had independently arrived at many of the same conclusions.*

⑲ *The mathematician was George Boole. The algebra is called "Boolean" algebra.*

⑳ *Babbage invented the first programmable computer.*

㉑ *The emperor was Julius Caesar. The calendar was, and is, called the Julian calendar, and the month named for him is July.*

㉒ *It was Sir Isaac Newton, whose law of universal gravitation allowed scientists to calculate the orbits of planets and satellites with great accuracy. Newton's suggestion for an artificial satellite was made in 1687, 270 years before the Soviet Union launched Sputnik in 1957.*

㉓ *Surprisingly, Arabic numerals as we write them were never used by Arabs. They were invented by mathematicians in India.*

㉔ *The odds are much better than even that, of six randomly selected people, two will have birthdays in the same month. In fact, the true odds are about four-to-one in favor of a shared birth month.*

㉕ *The "father of bookkeeping" was Luca Pacioli, an Italian who published an explanation of double entry bookkeeping in 1494.*

㉖ *The planet is Neptune. It was discovered by mathematical analysis of irregularities in the orbit of Uranus. Calculations pinpointed the location of a possible planet that could cause the irregularities. When astronomer John Galle turned his telescope on that portion of the sky, he found Neptune, discovering the new planet on the very first night he searched for it.*

㉗ *It has been calculated that there are more than 300 billion ways two players can make the first four moves of a chess game.*

㉘ *Many more than a billion. More like 806 million billion billion billion billion billion billion different arrangements of fifty-two playing cards are possible. You could shuffle a deck of cards for a lifetime—for many lifetimes—and the chances that the cards would ever repeat a sequence in its entirety are almost nonexistent.*

(29) *Arabic numerals were forbidden because it was thought that they could be forged or altered too easily.*

(30) *Ten plus ten equals 100 when you use binary arithmetic, the kind of math used by computers. In binary arithmetic, only the symbols "1" and "0" are used. In this number system "10" is the number we write as "2" in our base ten numbering system. Two plus two is, of course, four, and the way four is written in binary is "100." Thus, for computers, "10" plus "10" equals "100."*

(31) *It's the abacus, probably invented by the Chinese, but used also by early Greeks and Romans. This ancient device, which consists of a frame strung with wires on which beads can be moved up or down to carry out many mathematical calculations, is still in use today.*

(32) *It would most likely end in a tie. Both the abacus and the computer are limited by the speed at which the operators' fingers can move.*

(33) *The slide rule was invented by William Oughtred, an English clergyman, in 1621.*

(34) *Historians aren't sure about Euclid's birthplace but think it might have been*

Athens, Greece. More than 2,000 years ago he lived and worked in Alexandria, Egypt, then one of the largest, richest, and most sophisticated of the world's cities.

(35) *Blaise Pascal.*

(36) *ENIAC operated for the first time in February, 1946.*

(37) *It was the transistor, invented in 1947 by three researchers at Bell Laboratories. Their work gained John Bardeen, Walter Brattain, and William Shockley the Nobel Prize for physics.*

(38) *It is the Turing test, named for Alan Mathison Turing, a British pioneer in the field of artificial intelligence.*

A Better Mousetrap

It is said that where pessimists see problems optimists see opportunities. In no field of human endeavor is the truth of this idea more apparent than in the realm of inventions. Troubled by the scurrying of small rodents in your house? That's a problem. Figure out a better way to catch those mice—that's seizing an opportunity.

In the questions that follow, you have an opportunity to find out how much you know about some of the many inventors and the contraptions and techniques they have devised in an effort to solve some of the serious (and not so serious) problems of human existence.

1 Sir Hiram Maxim was a nineteenth century American inventor who was proud of some of his many inventions, which included a curling iron and a locomotive headlight. He was not proud of his most famous invention—the one that won him a knighthood in England. He later said, "I wish I had not invented it." What was the invention?

2 Today people can play any one of thousands of video games. What was the first successful home video game and when was it invented?

3 Two American inventors were working on the same invention at the same time. This invention—one of the most important ever made—changed the lives of everybody in the world. One of the inventors reached the patent office half an hour ahead of the other and is credited with this great invention. Who was he and what was his invention? Who was the other inventor?

4 Alfred Nobel, the Swedish inventor who is famous for inventing dynamite, also invented a building material that is used today in millions of structures all over the world. What building material did Nobel invent?

5 The granddaddy of all today's automobiles rolled along a street in France longer ago than you might think. What kind of a vehicle was it and who invented it?

6 A vital part of every automobile is the universal joint (which gives the drive shaft flexibility). It was not invented in Detroit. Who did invent it and when?

7 How and by whom was the word *robot* invented?

8 History books may tell you that the first electric battery was invented in 1800 by the Italian Alessandro Volta (after whom the electric unit known as a volt is named). But there is reason to believe that batteries were known long before Volta's time. An archaeological discovery made in 1936 may be the remains of an ancient battery. Where was it found?

9 There is a lot of argument about who was the first inventor to launch a rocket propelled by the burning of liquid fuel. Americans usually attribute the invention to Robert Goddard, a professor of physics. Goddard tested his first rocket in 1926, but strangely never told anyone of his experiment for ten years. In the meantime, other researchers, working independently, had also built liquid-propelled rockets. But who really designed the first liquid-propelled rocket?

10 How long does a U.S. patent last: (a) 12 years, (b) 17 years, or (c) 100 years?

11 How long did Edison's first incandescent light bulb burn?

12 Some important inventions seem quite simple, especially after the inventor has done the hard work of solving the problem. Such an invention was one made by a French doctor, Rene Laennec, in 1816. He was examining a young woman who showed symptoms of heart disease and wished to listen to her heart, but he was embarrassed to put his ear to her chest. Problem? No problem. In a few minutes Laennec came up with an invention that is standard equipment for doctors everywhere. What was his invention?

13 Which famous American inventor, who was granted more than 1,200 patents during his lifetime, had only three months of formal schooling?

14 In 1902 an American printer designed a device that changed summertime forever. What was it?

15 Before the development of electronic computers, programmable machines were run by the means of punched cards. When and in what industry were the first punched cards used to control machines?

16 Which was invented first, the gasoline-powered automobile or the rubber tire?

17 Probably no other invention is as important to modern civilization as is paper.

Paper made of a reed called papyrus was used by Egyptians thousands of years ago. In the first part of the eighteenth century, a French scientist, Rene de Reamur, developed an idea for a new kind of paper when he watched wasps at work. What was Reamur's idea?

18 Which was invented first, the cigarette lighter or the match?

19 What everyday contrivance is rarely used for its original purpose of fastening shoes?

20 An American inventor was given a U.S. patent for a means of making steamboats ride lighter in the water so they could navigate shallow water. It was the only patent the inventor ever applied for, but, although his invention was never used, the inventor became very famous. Why?

21 Every kind of thing used in our technological world had to be invented by someone, sometime. Consider the familiar traffic light. A policeman in Detroit, Michigan, designed a traffic light for automobiles in 1920. He used lights of three colors—green, yellow, and red—to control the flow of traffic through intersections. We still use these same colors in traffic lights today. Why were these colors chosen in the first place?

22 How did a walk at night when the inventor was blinded by the headlights of an oncoming car lead to a major invention of the twentieth century?

23 During the Civil War, a manufacturer in New York City offered a large cash prize to anyone who could invent a substitute for ivory, out of which billiard balls were carved. After seven years of experimentation, John Hyatt, a printer in New York State, developed a mixture of cellulose nitrate and camphor. Properly mixed and heated together, these ingredients formed the first man-made plastic. What was it called?

24 In 1902 an inventor named C. E. Johansson took out a British patent for what would be called "jo-blocks" (with "jo" pronounced like the man's name "Joe"). Jo-blocks were important to the spread of manufacturing worldwide. What are they?

25 The largest corporation in the world, IBM (International Business Machines), had its beginnings when a statistician named Herman Hollerith built a certain machine. What was the machine and how did it lead to today's sophisticated computers?

26 As the 1800s changed to the 1900s, an Italian physicist, Guglielmo Marconi, made a series of discoveries that changed our world forever. In 1895 Marconi succeeded in broadcasting a wireless (radio) signal from a transmitter to a distant receiver. In 1899 he sent the first wireless telegraph message between two countries—England and

France. In 1901 he sent the first radio message across the Atlantic Ocean—between Canada and England. At the time of this first transatlantic broadcast, scientists did not understand how radio waves could travel across such a distance, because it was known that radio waves travel in straight lines, and Canada and England are "out of sight" of each other around the curvature of the Earth. A British physicist would discover the explanation in 1902. A feature of Earth's atmosphere would be named after him. What is it in the atmosphere that makes it possible to send radio waves around the world?

27 Most of the fabric in the clothing we wear today is made from man-made fibers, which are cheaper than the natural fibers they replaced. Manufacture of synthetic fibers is a huge worldwide industry. What was the first synthetic textile fiber to be patented in the U.S.?

28 It was quite an entrance. Joseph Merlin had decided to amaze the world with his inventive wizardry. At a fancy dress ball in London, Merlin came through the front door playing a violin. As amazed spectators watched, the violin-playing Merlin traveled the length of the hall and smashed into a huge mirror, seriously injuring himself. What had Merlin invented but not perfected?

29 Sometimes a successful invention is just a matter of an alert person keeping his or her eyes open to some common occurrence. That's what happened to George de Mestral, a Swiss engineer hunting in the mountains. At first he was merely annoyed by the seed burrs that clung tenaciously to his clothing. Then he got to wondering about the burrs' amazing power to grab cloth and hold onto it. Close study showed that the burrs clung because they possessed small reverse barbs that latched over fibers in cloth (and hairs of passing animals). This realization gave de Mestral an idea for a very profitable, widely used invention. What is it?

30 A certain German immigrant to the United States, a man named Strauss, headed for California to find his fortune in the newly discovered goldfields. Strauss found wealth, all right, but he never found a gold mine. Instead, he started a profitable business making durable clothing for miners. The garment he developed is still worn and still bears his name. What is it?

31 Go in almost any modern building and you'll see the descendants of an invention made about 1860 by the French scientist Alexandre Becquerel. What was the invention?

32 Clarence Birdseye was a fur trader who went to Labrador, Canada, during World War II. While there he watched how the Eskimos stored the fish they caught. These observations later led him to make an important invention that surely has tickled your taste buds. Do you know what Birdseye's invention was?

33 In the 1890s two brothers, one a hospital administrator and the other the chief physician of the hospital, made an accidental discovery that led to the production of a new kind of breakfast food. What was the brothers' last name, and what industry did they pioneer?

34 Ancient peoples used various kinds of cement to construct buildings and roads. The best cement of antiquity was fashioned by the Romans, many of whose works remain intact today. Today we use a kind of cement called "Portland" in our construction projects. It was invented in 1824 by an English bricklayer who was looking for a better mortar. What is Portland cement made of?

35 One of the most inspiring and astonishing invention stories in history is that of a French fifteen-year-old who invented something that would enable millions of people to read. What was his name and his invention?

36 In northern Iraq thousands of years ago, people made pottery out of a kind of clay that contained a lot of a certain metal. Pure samples of the metal, however, were not obtained until 1825. For many years the metal was rare and precious, although it is commonplace today and is used in many inventions that make modern life possible. What is the metal?

37 In the last half of the nineteenth century, a growing network of railroads

covered the United States. Like all machines, the engines needed lubricating oil to keep them from getting too hot. Trains stopped frequently while crew members clambered around to squirt oil in places subject to friction. Elijah McCoy, who was born in Canada to parents who had run away from slavery in the American south, thought that this frequent stopping of the trains was an unnecessary inconvenience. He invented an automatic lubricator that oiled railroad engines while they were moving. The idea of the automatic lubricator caught on, and soon variations of McCoy's device were used on a variety of machines. McCoy's invention was a commercial success and also added a phrase to the language. What was the phrase?

38 In 1854 an English manufacturer, Henry Bessemer, invented a process that revolutionized technology. What is "the Bessemer process"?

39 What's fiberglas made of?

40 What time is it? Before the invention of clocks, people could answer this question only with such generalities as morning, mid-day, or evening. The first clocks were bulky and expensive and were typically placed in church towers, where they rang out the hours to citizens of the neighborhood. It was not until after the year 1400 that an Italian named Brunelleschi made an invention that made possible the manufacture of small clocks. What was Brunelleschi's invention?

41 In 1958 Richard Knerr and Arthur Melvin invented a simple toy that became one of the greatest marketing successes of all time. Consisting merely of a ring of plastic, they sold 40 million dollars worth of their toy in the first six months. What was this toy that caught on so quickly?

42 It is said that the kite was invented in the second century B.C. by a Chinese general named Han Sin. He used the kite for a military purpose. What was the miliary function of these first kites?

43 In the seventeenth century, the Dutch inventor Christian Huygens drew paper plans for a new kind of engine. It was to be powered by the force of a series of explosions of gunpowder. Though Huygens' engine was never built (and would never have worked), his invention is one of the most important in history. What was Huygen's idea?

44 For twelve hours, the new boat invented by Cornelius Drebbel navigated up and down the Thames River. Yet none of the many people on the banks of the river could see the boat. Why?

45 In what year were television sets first available to consumers?

46 Okay, so television is older than most people think. How about color television? When was color television developed?

47 Who really invented the first airplane—that is, a heavier-than-air flying machine?

48 Wilhelm Roentgen invented X-ray photography in 1895. How long did it take medical people to get around to using the new invention in medical diagnosis?

49 Modern communications systems use fiber optics to transmit messages by light. These systems only came into use in the 1970s, but actually the discovery that light can carry messages was made long before by a famous inventor. Who was he?

50 An invention made by Ben Franklin led to a fad in the kind of hats worn by women in London. What was the invention?

Answers

1 *It was the first successful machine gun capable of firing lots of bullets quickly. The "Maxim gun" was invented in 1884. The destructive power of this weapon and its modern descendants changed warfare forever.*

② *Noland Bushnel, an American engineer, invented the game "Pong" in 1972. Pong got its name from ping-pong. Players moved an electronic paddle up and down, blocking a video blip that moved back and forth. Today's video champions would find Pong too simple. Bushnel sold Atari, the company he formed, for $28,000,000 in 1976.*

③ *The inventor who won the race to the patent office was Alexander Graham Bell, and his invention was, of course, the telephone. The inventor who lost out by an unlucky half hour was Elisha Gray. Legal battles between the two men went on for years. In the end, Bell won the patent rights because of his earlier arrival at the patent office.*

④ *The invention was plywood: thin sheets of wood laminated (glued) together. By crisscrossing the grain of the wood in alternate layers, plywood is made much stronger than ordinary wood of the same thickness. It is believed that Nobel, appalled by the destructive power of the explosive he had invented, established the Nobel prizes—awarded to peacemakers, writers, and scientists.*

⑤ *The ultimate prototype of today's cars was a steam automobile. It achieved a top speed of just a little more than two miles per hour (you can walk twice as fast!). The inventor was Nicolas Cugnot, and he demonstrated his invention in the year 1771!*

⑥ *The universal joint was invented long before the first automobile was built. It was described in 1545 by the Italian Geronimo Cardano.*

⑦ *The word* robot *was coined by the Czechoslovakian playwright Karel Capek in 1921. The word was formed by dropping the ending of* robota, *the Czechoslovakian word for work.*

⑧ *It was found in present-day Iraq. The discovery consisted of a clay pot containing a copper cylinder and an iron rod coated with lead. Were such a device filled with a proper liquid (either an acid or an alkaline solution), it would generate an electrical current of less than one volt. If this archaeological mystery was indeed a battery, then human's knowledge of electricity goes back at least 2,000 years!*

⑨ *Many historians of science believe the first liquid-propelled rocket device was the one patented by the Peruvian engineer, Pedro Paulet, in 1895.*

⑩ *The correct answer is (c). A U.S. patent is granted for seventeen years.*

⑪ *Edison's first successful light bulb burned for forty hours before the filament burned out. Today a standard twenty-five-watt commercial light bulb lasts an average of 2,500 hours before you have to buy a new one.*

⑫ *The invention was the stethoscope, which enabled Laennec to listen to his patient's heart without immodest contact.*

⑬ *The self-taught inventor was Thomas Edison. His first and only schoolteacher thought he was crazy. Edison dropped out of school at the age of seven and never returned.*

⑭ *The invention was air conditioning. It was developed by Willis Carrier. Carrier was troubled by printing problems during hot weather. He concocted a machine that could cool a room by blowing air over refrigerated pipes.*

⑮ *The idea of controlling machinery with punched paper was developed in France in the early part of the nineteenth century. This technique was pioneered in the textile industry to control the weaving of patterns into fabric.*

⑯ *If you guessed that the rubber tire came first, you're right. A Scottish engineer, Robert Thomson, came up with the idea in 1845. Now guess what kind of vehicle was the first to ride gently on soft rubber tires? The answer is the bicycle.*

⑰ *Reamur's idea was that the paper could be made out of wood. He got the idea after watching wasps chew wood and spit it out into a paper paste they used to build their nests.*

⑱ *The correct answer depends on what you mean by a match. One kind of match, made of wooden sticks that were dipped into a sulfur mixture, was invented in 1681 by Robert Boyle. They could only be lit by placing the head in contact with fire. The ancestor of the cigarette lighter was invented by a German chemist, J. W. Dobereiner, in 1816. The first modern match was made in 1844 by a Swedish inventor, Carl Lundstrom.*

⑲ *Whitcomb Judson, an American inventor, patented what he called a "Clasp Locker and Unlocker for Shoes." Today's improvements of Judson's invention are called zippers. Coats, pants, handbags—these items and more come with zippers; shoes almost never do.*

⑳ *Because the inventor was Abraham Lincoln, who later became president of the United States.*

㉑ *William Potts, the originator of the modern traffic light, chose the colors green, yellow, and red because they were used by railroads. Their meanings were already known: green = go; yellow = slow; red = stop.*

㉒ *Blinded by the headlights, a college student named Edwin Land got to wondering if he could invent a lens that would polarize light and reduce glare. He patented a kind of lens still used in sunglasses and cameras today. But it was his invention of the Polaroid camera for which he is most famous.*

(23) *The name of this first synthetic plastic is celluloid. Though it is said that celluloid did not make a very good billiard ball, celluloid was an important invention. It made possible the development of the film used in modern still and movie cameras.*

(24) *Johansson realized there was need for a more accurate method of measuring machine parts. His invention was steel blocks that were milled to highly precise measurements. Their use made it easier for industry to make parts the same size no matter where in the world the parts were made.*

(25) *Hollerith built a punch card information analyzing machine to break down information accumulated in the 1890 U.S. census. Today's electronic computers do the same kind of work (and much more) at much greater speeds.*

(26) *It was discovered that a layer of the atmosphere containing a large number of electrically charged particles reflects radio waves. In effect it acts like a radio "mirror," making it possible to "bounce" radio waves around the curvature of the Earth. This atmospheric feature was named the "Heaviside layer" after Oliver Heaviside, its British discoverer.*

(27) *The first synthetic textile fiber was rayon, patented in 1902.*

(28) *Roller skates. Unfortunately for Merlin, his design did not allow the skater to*

steer, thus his accident. It was during the Civil War, in 1863, that a Massachusetts inventor, James Plimpton, hit on the idea of placing flexible cushions between the foot plate of the skates and their wheels. This flexibility made it possible for skaters to shift direction by shifting their weight.

(29) *The barbed burrs gave de Mestral the thought that maybe he could invent a fastening device that worked on the same principle. His invention is known under the tradmark "Velcro." One side of a Velcro fastener consists of small hooks of plastic; the other consists of loops.*

(30) *Strauss designed a kind of pants made out of cotton denim with reinforcing rivets at points of stress. Pants built to much the same design are worn today and are known as Levi's. Levi was Strauss's first name.*

(31) *We think of fluorescent lighting as being very modern and high-tech, but the basic idea was worked out by Becquerel more than 130 years ago. He coated a glass tube on the inside with a fluorescent substance that glowed when electricity passed through the tube.*

(32) *Birdseye's invention was the first successful process for freezing and storing food products. The secret of his success was using very cold dry air, just like the naturally occurring cold air of the Arctic.*

(33) *The brothers were Will and Dr. John Kellogg. They had been trying for a long time to find a way of processing wheat into ready-to-eat form. They boiled wheat and tried to flatten it, but the boiled wheat just formed a sticky mess that gummed up the rollers. The accident that led to their discovery came about when they forgot a batch of boiled wheat for a few days. To their surprise, this batch rolled out into individual flakes that were delicious when toasted. The Kelloggs developed their wheat flake cereal food and pioneered the breakfast cereal industry.*

(34) *Portland cement is made of crushed limestone and clay. The powdered mixture is roasted at a high temperature to remove almost all the water in the material.*

(35) *The inventor was Louis Braille, a blind teenager. He conceived the idea of representing letters by raised dots on paper so that blind persons could read with their fingers.*

(36) *The metal is aluminum. It was not until mankind developed large powerful sources of electricity that it was possible to extract aluminum from its ore.*

(37) *McCoy's automatic lubricators worked so well that purchasers insisted on buying those made by his company—demanding "the real McCoy."*

(38) *Bessemer invented the modern method of turning iron into steel.*

(39) *If you thought this question was supposed to be tricky, you probably got the wrong answer. Fiberglas, just like its name says, is made of glass. It is formed of fibers of glass that have been "drawn out" while the glass is hot and liquid.*

(40) *Brunelleschi invented the spring-driven clock. This made it possible for the first time for people to have clocks in their homes.*

(41) *It was the Hula-hoop.*

(42) *The first kites were used to send messages. Since high-flying kites were visible over a great distance, kites of different colors and shapes could be used to transmit information to troops separated from one another.*

(43) *Gunpowder would never have worked as a fuel, but Huygens had worked out all the basic principles of the internal combustion engine, in which a controlled series of explosions in an engine drive pistons that in turn drive a crankshaft. Today, of course, descendants of Huygens' engine, powered by gasoline, power almost all the automobiles in the world.*

(44) *No, Drebbel had not invented "stealth" technology. The year was 1624, and Drebbel's contribution to shipbuilding was the invention of the first submarine.*

(45) *Most people are surprised to learn that television sets were on the market in 1930. They were designed and built by John Baird, a pioneer in the development of television (which means the sending of images across a distance).*

(46) *Again, the answer may surprise you. John Baird built the first crude color television system in 1928.*

(47) *It wasn't the Wright brothers, who were the first to fly an engine-powered aircraft in 1903. Nearly a century before that, in 1809, an Englishman, Sir George Cayley, constructed and flew a glider.*

(48) *Not very long. Less than two months after Roentgen announced his great discovery, it was used to study the brain of a patient.*

(49) *Alexander Graham Bell. He invented what he called the "Photophone" in 1880, almost a century before experimenters put the discovery to use.*

(50) *The lightning rod. Women in the late eighteenth century wore lightning rods on their hats and trailed ground wires behind them to conduct away the electrical charge should they happen to be struck by lightning.*

Man, The Builder

Humankind has been described in many ways, but one thing is certain, man is a builder. Throughout history people have designed and built bridges, buildings, roads, tunnels, dams, mines, forts, towers, arches, pyramids, domes, and myriad other kinds of structures.

Engineering, architecture, and construction technology are hardly mere matters of wood, stone, masonry, glass, metal, earth, and space-age plastics. Every one of the countless building projects ever carried out by men began as an idea, and every successful project completion is a testimony to the power of the human intellect.

Find out how well your intellect is functioning as you answer these questions about man, the builder.

❶ What were the first construction projects ever carried out by human beings?

❷ The development of agriculture was necessary before humans could live in fixed abodes. To protect themselves and their stored crops from the depredations of human raiders, many early agriculturalists soon found they had to build fortifications around their dwelling places. Walled towns may have been the earliest large-scale engineering projects. Can you name the oldest known example of defensive walls built around a town?

❸ An ancient mine has been discovered in the Netherlands, and it has been determined (by radiocarbon dating) that humans worked this mine at least 5,000 years ago. The mine covers sixty-two acres and contains separate shafts. Because this remarkable mine existed before the age of metals, what prompted the Europeans of long ago to delve so assiduously into the earth?

❹ What was the largest pyramid ever built?

❺ The pyramids in Egypt have stood for thousands of years. What structure in the U.S. is likely to last that long or longer? You'll probably guess that maybe it's one of the big dams or a famous skyscraper, but you'd be wrong. Nobody could accuse you of ignorance if you don't get the right answer, because the structure that will last longest is one you've never seen, though you may have heard of it. What is it and where is it located?

6 The Grand Coulee dam in Washington State is the largest concrete structure in the world. In terms of mass, how does its bulk compare with that of the largest Egyptian pyramid?

7 The Tower of Babel, mentioned in the Bible, is one of the most famous structures ever built. What did the ancient Babylonians call such a structure?

8 What color was the Tower of Babel?

9 Here are seven structures built long ago, each of which was an engineering achievement of its time: The Temple of Artemis, the Colossus of Rhodes, the Lighthouse on Pharos, the Great Pyramid at Giza, the Statue of Zeus, the Tomb of Mausolus, and the Hanging Gardens of Babylon. What collective label do we use for these structures?

10 More than two thousand years ago Roman engineers invented a machine of a kind still used today in constructing the underpinnings of bridges and buildings. What was their invention?

11 The Roman emperor Caligula ordered his military engineers to build what for centuries would hold the record as the world's longest bridge. It stretched across the bays of Baia and Puteoli south of Rome. Not wanting to be thrown to lions if they failed, the engineers quickly solved the problem of building the three-and-one-half-mile long bridge. What kind of a bridge did they build?

12 One of the most amazing engineering achievements of the ancient world was the method used by builders of the Roman Empire to bring supplies of fresh water to their cities. How did they do it?

13 Can you name the military empire that constructed the first system of well-engineered roads?

14 The greatest road builders of ancient times were, of course, the Romans. In 200 A.D., the network of Roman roads reached from the present-day border of Scotland and England in the northwest to the Euphrates River in the southeast of the empire. A first-class Roman road was paved, and consisted of four layers of material. What were they?

15 About two centuries B.C., a Chinese emperior, Shih Huang-ti, assumed rule of China. During his reign, he linked together a series of fortifications on the empire's northern borders, forming the engineering marvel we know as the Great Wall of China. How was the Great Wall constructed?

16 Temples and churches have always been among the most ambitious construction projects undertaken by men. Large buildings built before 1200 A.D. had few windows in their walls. After that, European churches were built with interiors that were splendidly lit by huge stained-glass windows. What architectural development made possible this new kind of building?

17 The church builders of Europe vied with each other to build the tallest spires on their churches. How high was the tallest church spire built in Europe?

18 What is the tallest building on Earth today? How high is it?

19 What's the tallest structure ever built by mankind?

20 The tallest monument in the world is an arch made of stainless steel. It is 630 feet high and is as wide as it is high. What is this monument and where is it located?

21 A famous American architect designed a mile-high building. Who was the architect, and did he really believe that such a structure could be built?

22 What famous skyscraper was designed to serve as a mooring mast for dirigibles?

23 People on the top floors of the World Trade Center in New York City (1,362 feet,

3¼ inches tall), say they feel queasy and become airsick when a strong wind is blowing. Is it just their imaginations, or do really tall buildings actually sway so much that people can be affected by the movement?

24 Where is the oldest surviving bridge ever built?

25 The two huge towers that support the world's longest-span suspension bridge, which is over the Humber River in England, were constructed to lean away from each other 1.44 inches. Why did the engineers design them that way?

26 The world's widest long-span bridge carries two railroad tracks, eight automotive vehicle lanes, a bicycle path, and a sidewalk. Where is this bridge located?

27 A famous bridge was moved from its former location. What was the bridge, and where did it end up?

28 When London Bridge was built (near the end of the twelfth century), a workforce of 800 men took more than seven years to complete the construction. To get some idea of the rate of technological advance, consider this question: How many men do you think worked for how long to rebuild it in Arizona?

29 A bridge that became famous because it fell down was the Tacoma Narrows Bridge in the state of Washington. This 2,800-foot span across Puget Sound was nicknamed "Galloping Gertie" soon after it was built. It swayed in the wind, often moving as much as thirty feet in a horizontal direction. Perhaps the builders should have been worried, but their calculations convinced them that the bridge was structurally sound. After all, they had designed it to withstand wind gusts of 120 miles per hour. Yet one day, in a mere forty-two-mile-per-hour wind, the bridge began to undulate in ever-widening oscillations until it snapped and collapsed. What did the engineers do wrong?

30 How long have humans been building canals to control the flow of water?

31 It was no doubt one of the most ambitious and labor-intensive construction projects in history, lasting from 540 B.C. until 1327 A.D. At one time, it is estimated that 5 million laborers were working on it at one time. What was this mammoth engineering endeavor?

32 Where is the longest man-made waterway in the world?

33 The fees charged to vessels going through the Panama Canal are based on their tonnage. The largest fee ever paid was $89,154.62, by the liner *Queen Mary*. The smallest fee ever charged was thirty-six cents. What kind of a ship paid such a tiny amount?

34 The construction of tunnels is the most dangerous engineering endeavor. In terms of human death and injury, what is the most costly tunnel ever built?

35 In 1876 a visionary Frenchman died in poverty, his life-long dream an object of ridicule. His name was Thome de Gamond. What was his visionary project and why do we hear so much about it today?

36 When and where were the first deep wells drilled for the purpose of obtaining natural gas?

37 The trans-Alaskan pipeline runs 800 miles across Alaska. How many rivers and streams does it cross?

38 Why was the trans-Alaskan pipeline built on the surface, instead of being buried in the ground as most pipelines are?

39 The biggest man-made hole in the world is a mine, of course. It is a mile wide and 4,000 feet deep. Twenty-five million tons of rock and earth were removed from it. What kind of a mine was it?

40 What is the strongest shape that can be used in construction?

41 Buckminster Fuller, the inventor of the geodesic dome, once designed a dome

large enough to cover a major American city. What city was it?

Answers

① *Nobody knows for sure. But the building of shelter from the elements must have been early on the list of man's engineering priorities. The oldest-known evidence of human building activities is a piled circle of blocks of volcanic rock found in Olduvai Gorge in Tanzania, Africa. It is believed that a band of our ancestors put these stones in place about 1,700,000 B.C.*

② *This distinction belongs to the famous Biblical city of Jericho, located in a few miles north of the Dead Sea. In this desert oasis, farmers built walls around their town as long ago as 8350 B.C.*

③ *The ancient miners were digging for flint to make heads for their axes. Scientists calculate that during the lifespan of the mine enough flint was extracted to make an astounding 150 million axe heads.*

④ *Most people get the answer to this question wrong, picking the great pyramid of Cheops in Egypt. The most massive pyramid on Earth, however, is in Mexico. It is the Pyramid of Quetzlcoatl, which contains 4,360,000 cubic yards of material. The pyramid of Cheops contains a mere 3,360,000 cubic yards.*

⑤ *It's the North American Aerospace Defense Command facility near Pike's Peak in Colorado. The steel buildings there are deep inside Cheyenne Mountain, with a 1,700-foot-thick granite roof protecting them from possible nuclear attack and the ravages of time.*

⑥ *Grand Coulee dam contains more than 10.5 million cubic yards of concrete and is therefore more than twice as massive as the pyramid of Cheops.*

⑦ *They called it a "ziggurat," from a word in their language meaning "to build high." The Tower of Babel may have been as much as 200 feet in height (about as tall as a modern sixteen-story building).*

⑧ *Built by Nebuchadnezzar in the sixth century* B.C., *the structure that would become known as the Tower of Babel consisted of seven different levels, each a different color. From bottom to top, the levels were white, black, blue, yellow, silver, and gold.*

⑨ *They are known as the Seven Wonders of the Ancient World. Only one of them is still standing—the Egyptian Pyramid at Giza. It is 4,900 years old.*

⑩ *It was a pile driver. The Roman device consisted of a huge stone that was first lifted by a giant wheel and then dropped onto the top of a post to drive it into the ground.*

⑪ *A pontoon bridge. The Roman engineers laid planking across two rows of ships.*

⑫ *They built stone conduits (channels) that carried water from high mountain lakes to the cities. These structures were called aqueducts. They tunneled through mountains and crossed valleys on a series of stone-built arches. The longest was the Aqueduct of Carthage in Tunisia, which carried water more than eighty-seven miles.*

⑬ *It was the Assyrian Empire of Mesopotamia, whose road building began about 1000 B.C. The Assyrian roads were not paved, but were leveled tracks that carried royal couriers and chariots of war.*

⑭ *A first-class Roman road began with a layer of sand, on which were successively placed a layer of rubble and a layer of concrete. The surface was paved with flat stones cut to fit closely together. These roads allowed swift travel in the empire. History records that, traveling on these roads, Julius Caesar journeyed from England to Rome in a mere six days' time.*

⑮ *The Great Wall of China was formed of layers of rammed earth faced with brick or plaster. Built as a defense against "barbarian" Asian tribes, the Great Wall in some sections rises to a height of nearly forty feet, is about twenty-five feet wide at the top, and runs for a continous distance of 1,400 miles.*

⑯ *It was the development of the so-called "Gothic" arch (which was actually based on Islamic arches seen by Crusaders). The Gothic, or pointed, arch replaced the round-topped "Roman" arch. In Roman-arch construction, the walls must be massive, to support the weight of the roof. The Gothic arch carries the roof load more efficiently, and allows for the use of more windows in architectural design.*

⑰ *The spire of the Lincoln Cathedral in England was completed in 1548. It towered well over 500 feet above the ground. Perhaps the builders were too daring; the spire collapsed during a violent storm.*

⑱ *The tallest building on Earth is the Sears Tower in Chicago. Its 110 stories reach 1,454 feet above the streets below. The Sears Tower was completed in 1974.*

⑲ *The tallest structure ever built is a radio-broadcast mast near Konstantynow, Poland. Completed in 1974, this structure is 2,120 feet and 8 inches in height. Its tubular-steel column is held upright by fifteen steel guy ropes that anchor it to the ground.*

⑳ *The world's tallest monument, dedicated to the pioneers who risked their lives in opening up the American West, is the Gateway Arch to the West. It is in St. Louis, Missouri, at the site of what was once called Westport Landing, the take-off point for the bold covered wagon journeys of the nineteenth century.*

㉑ *The architect was Frank Lloyd Wright. Many engineers have confirmed the soundness of his design. This ultimate skyscraper has never been built, of course, and nobody knows whether Wright was serious about his plans.*

㉒ *The Empire State Building in New York City.*

㉓ *Astonishing as it may seem, the twin towers of the World Trade Center, in New York City, can each move as much as three feet in a strong wind, and as much as seven feet to one side or another in a wind of hurricane force. Looking up from the ground, you can't see the swaying, but some people really notice it when they're way up there on one of the top floors.*

㉔ *It crosses the River Meles in Izmir, Turkey, and dates from 850 B.C. It consists of a single arch of stonework.*

(25) *The Humber River span is 4,626 feet in length, fourteen times as long as an American football field. The towers are so far apart that it was actually necessary to allow for the curvature of the Earth.*

(26) *This wide long-span bridge is located in Sydney, Australia.*

(27) *London Bridge, which once spanned the Thames River in England, was moved, stone by stone, to Lake Havasu City, in Arizona.*

(28) *With the assistance of modern twentieth century machinery, it took forty men only two years to put London Bridge back together again.*

(29) *The designers of the bridge slipped up by not allowing for a steady wind that would keep the bridge swaying back and forth in what scientists call harmonic rhythm, until it moved so far that its suspension cables could no longer take the strain.*

(30) *Archaeologists working in the Middle East have found evidence of canal building dating from about 4000 B.C. These canals were constructed to carry water to agricultural crops.*

(31) *It was China's Grand Canal, which when finished, stretched more than 1,100*

miles. It served the twin purposes of helping the Chinese control the drainage of rivers they depended upon for irrigation and transporting people and cargo.

㉜ *In the former Soviet Union. The Volga-Baltic waterway, 1,850 miles long, runs from the mouth of the Volga River on the Caspian Sea to the city of St. Petersburg on the Gulf of Finland.*

㉝ *It wasn't a ship at all. The thirty-six cent fee was charged to a swimmer, Richard Haliburton, a travel writer and adventurer who swam through the Panama Canal in 1928.*

㉞ *The St. Gotthard Tunnel through the Swiss Alps, completed in 1882, holds this dismal record. Construction of this tunnel took a total of 310 lives, and 877 workers suffered permanent disabling injuries.*

㉟ *De Gamond dreamed of the building of a tunnel beneath the English Channel. He exhausted his own considerable fortune while testing the geology of the seabed and continued his research while living on the earnings of his daughter, a music teacher. More than a century after de Gamond died, the English and French governments agreed to build a tunnel such as he had envisioned, a thirty-mile-long tunnel between Calais, France, and Dover, England, popularly dubbed the "chunnel."*

㊱ *Would you believe 2,000 years ago, in China? This fact amazes many of us, conditioned as we are to thinking that our western, scientific culture has always been at the forefront of technology. Using human labor as motive power, the ancient Chinese were able to drill wells nearly 5,000 feet deep through solid rock. They used cast iron drill bits suspended from bamboo cables. Men jumped on and off a long lever attached to the cables, thus lifting and dropping the drill bit, pulverizing the rock. Broken rock was sucked out of the hole through long bamboo tubes. Some of the Chinese wells took many years to drill. Europeans learned how to drill deep wells only after the Chinese methods were described by a French missionary in the 1820s.*

(37) *Engineers building the trans-Alaskan pipeline had to cope with the problems of building it across as many watercourses as there were miles—800.*

(38) *The builders did not bury the pipeline because of the permanently frozen soil, or permafrost. They calculated that if they did bury the pipe, its heat would melt the soil around it, making it sink to a depth of thirty feet in a ten-year period.*

(39) *It was a diamond mine, known as "The Big Hole of Kimberly," in South Africa. It was operated from 1871 to 1915. Unlike today's diamond mines, in which heavy machinery is used, most of this mine was dug by workers using picks and shovels.*

(40) *It is the triangle, the basic structural unit of many building designs, one of the newest being that of geodesic domes pioneered by the American engineer, Buckminster Fuller.*

(41) *New York City. Will it ever be built? Most likely not. But so strong is the basic design concept that engineers are confident the structure would work.*

The Sky and Space Adventure

People must have dreamed of flying ever since the earliest humans looked in wonder at birds soaring in the wide sky.

The first story of human flight with the aid of a technological device is told in the Greek myth of Daedelus and Icarus. Daedelus, who was the inventor of many marvelous contrivances, worked for a powerful king who ruled the island of Crete. Daedelus wished to leave the service of the king but was held prisoner. He constructed, out of feathers held together with wax, two sets of artificial wings—one set for himself, another for his son, Icarus. Before their escape, Daedelus warned Icarus not to fly too high, lest he get too close to the sun. But Icarus became so intoxicated with the thrill of flying that he soared higher and higher. The heat of the sun melted the wax in his wings and Icarus plunged to his death in the sea.

Sadly, the history of aviation is replete with real-life tales of those who died because of mechanical failure. But it is also rich in stories of triumph, as humans have learned to fly higher, farther, and faster until escaping Earth's atmosphere entirely to launch ventures into the vastness of space.

See how many of these questions about our aerospace achievements you can answer.

1 Watching the flight of birds, many early aviation theorists believed that the best way to construct a flying machine was to construct a device with wings that flapped up and down like those

of feathered flyers. Can you name the English thinker who, in 1809, was the first to propose sound scientific arguments for building heavier-than-air craft with "fixed" wings?

2 In 1842 an Englishman, William Henson, developed plans for a powered aircraft that he dreamed would be able to carry passengers between the capital cities of Europe. To finance his dreams, Henson attempted to sell stock in a company he called the Aerial Transit Company. It was the first airline, but it never made a single flight. Henson's aeronautical design had incorporated a fatal flaw. What was it?

3 In December, 1903, just nine days before the Wright brothers' epochal flight at Kitty Hawk, North Carolina, an esteemed scientist, Samuel Pierpont Langley, director of the Smithsonian Institution, stood on a boat moored in the icy waters of the Potomac River, near Washington, D.C. His new and improved "aerodrome" (that's what he called his flying machine) was being catapulted skyward. The flight was a disaster. Langley's aerodrome somersaulted once and crashed in a tangle of wreckage. Langley was perplexed and astounded; his calculations had told him his craft was capable of flight. Did Langley's aerodrome ever fly?

4 Early aircraft were square and bulky contraptions. In 1912, however, a French pilot, Jules Vedrines, advanced the science of

aeronautics when he became the first aviator to break the 100-mph mark. What basic concept of aircraft design was proved workable by the airplane Vedrines flew?

5 Men soared above the earth long before the development of airplanes. The Montgolfier brothers of France were building successful balloons as early as 1783. What scientific principle explains the success of the Montgolfiers' balloons?

6 Who were the first passengers to ascend in a balloon?

7 The first humans to ride the air beneath a hot-air balloon were Pilatre de Rozier and the Marquis d'Arlandes, on November 21, 1783. How high did they fly? How far did they travel?

8 Hot-air balloons require fire, of course, and fire requires fuel. What fuel did de Rozier and the Marquis d'Arlandes use in their triumphant balloon ride?

9 One of the limitations of ballooning is that, unless it has a propellor like a dirigible or a modern blimp, a balloon is at the mercy of the winds. Its passengers cannot control the direction of their flight. In November, 1784, a French balloonist, Jean-Pierre Blanchard, made a flight in England in a balloon equipped with oars, to row through the air. Did the oars work?

10 Balloons were used during the American Civil War, in 1862. What military purpose did they serve?

11 In 1862 the Balloon Committee of the British Association for the Advancement of Science commissioned two men to attempt ascent to an altitude of 30,000 feet to take meteorological readings. The two passengers, James Glaisher and Henry Coxwell, actually achieved an altitude of 35,000 feet, but they almost lost their lives in the attempt. Why?

12 Through the years, balloonists tried to achieve ever and ever higher altitudes. The 50,000-foot level was exceeded in 1931, the 100,000-foot level in 1957, and the current world record was set in 1961. How high is this record and why is it unlikely that it will ever be significantly exceeded?

13 What was the first aircraft ever to carry passengers on a round-the-world journey? When did this flight take place?

14 What is the difference between a dirigible and a blimp?

15 Early in the development of airplances, explorers realized that this new method of travel offered great opportunities for penetrating inhospitable areas of the Earth. Naturally, adventurous explorers raced to be the first to reach the North or South Poles of the Earth by airplane (even

though these locations had already been reached by ground travel). Which Pole was flown to first, the North or South? Were these two polar aviation firsts achieved by the same person or by different people?

16 In what year did Captain Charles Yeager become the first pilot to fly faster than the speed of sound?

17 Which was invented first, the airplane or the parachute?

18 In 1957 the Earth acquired a new moonlet, which weighted 185 pounds. Where did the moonlet come from?

19 Can you name the first space traveler from Earth to die in space?

20 Two men, one American, the other Russian, are credited with being the pioneers of modern rocketry. Can you name them?

21 Most students of spaceflight know that Neil Armstrong was the first human to set foot on the moon, in July, 1969. But not too many people can name the two other astronauts who accompanied him on that famous *Apollo 11* flight. Can you?

22 Who was the first astronaut to be photographed standing on the lunar surface?

23 A device called the Lunar Rover Vehicle (LRV) was used by astronauts of the *Apollo 15, 16,* and *17* lunar missions. Powered by silver-zinc batteries, this electric "car" enabled the astronauts to explore portions of the lunar surface. How fast did it travel?

24 The astronauts of the *Apollo* missions brought back more than 2,000 rock and soil samples from Earth's only satellite. These samples have been distributed to laboratories and museums all over the world. Taken all together, how much do they weigh?

25 The National Aeronautics and Space Administration (NASA) is the U.S. Government agency in charge of spaceflight. Here's a three-part question about NASA: In what city is NASA headquarters located? In what city is the command center (ground control) for space missions located? Where is the primary launch site for American space missions?

26 Extravehicular activity (abbreviated EVA) is the space engineer's jargon for any activity carried out by humans in space outside the protection of their spacecraft, capsule, or space station. Who was the first American astronaut to carry out an EVA? When did he do it?

27 Is there life on Mars? So far, the answer seems to be no. Can you name the first space vehicle to make a close fly-by of the red planet?

28 Can you name the first space vehicle that actually landed on Mars?

29 After a disappointing series of four straight failures, the *Pioneer* series of American spaceflights achieved a number of space firsts which greatly extended our knowledge of the universe we live in. One notable space first was put into the record books by *Pioneer 4*, another by *Pioneer 10*. Give yourself full credit if you know what either one of these missions is noted for.

30 *Pioneer* and *Voyager* missions to Jupiter recorded two kinds of electromagnetic phenomena in its swirling hydrogen atmosphere. These also occur on Earth. What are they?

31 In 1980 an astronaut became the long-distance travel champion of all times as he completed his second long mission in space. What was his nationality? What was his name? How far had he traveled?

32 The space probe *Ulysses* was launched in 1990. Though its primary mission was to investigate the sun, it was first sent on a fly-by of Jupiter. Why did *Ulysses* take the 617-million-mile "detour" this route involved?

33 In November, 1974, scientists at the Arecibo Observatory in Puerto Rico beamed a powerful radio signal into space. Consisting of 1,679 binary characters, this message was sent in the hope that it might be received 25,000 years in the future. Right now it is traveling through space at the speed of light. What is the basic import of the message, and whom do we hope will get it?

34 Primitive organisms such as algae, bacteria, and fungi have been found living inside rocks on Earth's bleakest continent, Antarctica. What does the finding of these organisms have to do with the search for life elsewhere in the universe?

35 Suppose space engineers of the future built a space ship capable of hurtling through space at 99 percent the speed of light. Since this technological capability is far beyond that of the present day, let's suppose that the ship, with a single astronaut aboard, takes off in the year 2100, journeys at its design speed through space, and lands back on Earth forty years later, in the year 2140. Though forty years have passed on Earth, the clocks on the spaceship have measured off only a little more than five-and-one-half years, and the astronaut aboard has aged only that same five-and-one-half years. Why is this so?

36 Scientists have a name for a kind of particle that has never been observed and that may not even exist. It's called a tachyon, and if tachyons do exist they may make possible rapid interstellar travel and communications. How fast does a tachyon move?

37 Because of our upright posture, gravity shrinks us every day as the disks in the spine are squeezed. We regain the lost height every time we sleep, as the disks expand again. In prolonged weightlessness, such as that encountered by astronauts in space, height can change dramatically. How much taller are astronauts upon their return to Earth?

38 Cosmic rays pose a hazard both to astronauts and to the electronics systems of their spacecraft. What makes them so dangerous?

39 What is it that an exobiologist studies?

40 What is meant by the Fermi Paradox?

41 What is an IFO?

42 What is Project Cyclops?

Answers

① *He was Sir George Cayley. Cayley actually built and flew a glider. His success stimulated many other inventors to experiment with flying machines.*

② *Henson's plans envisioned an aircraft powered by a steam engine. Experiments proved, however, that the weight of steam engines of his time was so great that it was impossible to construct an airplane with enough lift to get the engine off the ground.*

③ *Yes. In 1914 Langley's aerodrome flew successfully. So even though he had not built the first engine-powered craft to fly, Langley had built the first machine that was capable of powered flight.*

④ *It was the idea of streamlining. Fashioning the shape of an aircraft to minimize the drag of air over the craft and provide lift makes faster speeds possible.*

⑤ *The Montgolfiers' first balloons were filled with hot air, taking advantage of the fact that a heated gas (air) is lighter than the same gas at a lower temperature.*

⑥ *They were a rooster, a sheep, and a duck. They rode in a wicker basket attached to a Montgolfier balloon, on September 19, 1784. The flight was witnessed by King Louis XVI and his court and lifted off from the elegant grounds of the palace at Versailles.*

⑦ *De Rozier and the Marquis d'Arlandes achieved an altitude of approximately 3,000 feet and traveled a horizontal distance of eight miles.*

⑧ *They fed the fire that kept their balloon aloft with bundles of straw thrown into an iron grate placed below the mouth of the balloon.*

⑨ *No.*

⑩ *Since a man in a balloon could see a lot more territory than one on the ground, balloon ascents were made to ascertain the position of enemy troops.*

⑪ *The two men nearly died of cold and oxygen deprivation. Until that time, no one had realized how hostile to earthbound life-forms were the upper reaches of our planet's atmosphere.*

⑫ *The official record altitude achieved by a balloonist was reached by Commander Malcolm D. Ross and Lt. Commander Victor A. Prather on May 4, 1961. Launching from the deck of a U.S. aircraft carrier, the two men attained a height of 113,740 feet. Though this is the official balloon altitude record, in 1966, an American, Nicholas Piantanida, reached an unofficial altitude of 123,800 feet. Piantanida died in a crash upon landing. It is unlikely that a balloonist will ever go much higher than Piantanida because that altitude is at the very edge of the Earth's atmosphere. Beyond is the airless void of space, where no balloon can ever travel.*

⑬ *The first aircraft ever to carry passengers around the world was the dirigible,* Graf Zeppelin, *in August of 1929. The 19,000-mile journey took twenty-one days, seven hours, and thirty-four minutes.*

⑭ *Dirigibles and blimps are both examples of what are called airships—that is, they are lighter-than-air aircraft that have steering and propulsion systems. The difference between the two is that blimps are inflated bags, whereas dirigibles have a rigid framework.*

⑮ *The North Pole was first overflown by Richard Byrd (acting as navigator) and Floyd Bennett (pilot) in May, 1926. Three-and-one-half years later, in November, 1929, Byrd (again acting as navigator) became one of the first two men to fly over the South Pole, but the airplane had a different pilot, Bernt Balchen.*

⑯ *The sound barrier was broken by Captain Yeager in 1947.*

⑰ *The parachute. The first successful parachute was demonstrated in 1783 by a daring Frenchman, Sebastien Lenormand, who leaped safely from a tower in southern France.*

⑱ *It came from what was then known as the Soviet Union. It was* Sputnik 1. (Sputnik *means* little moon *in Russian.) It was, of course, the first man-made object to be placed in Earth's orbit.*

⑲ *The first space casualty was the dog, Laika, sent up in a Soviet launch in 1957. The spacecraft was not designed to return to Earth. A week after launch, Laika died because oxygen aboard the spacecraft was depleted.*

⑳ *The American was Robert Goddard; the Russian was Konstantin Tsiolkovsky.*

(21) *They were Michael Collins and Edwin Aldrin, Jr.*

(22) *It was Edwin ("Buzz") Aldrin. Armstrong, who had disembarked first, took his picture.*

(23) *Astronauts of the* Apollo 17 *mission "put the pedal to the metal" and topped out a speed of about eleven miles per hour.*

(24) *Weighed under Earth gravity, the lunar samples brought back by the Apollo astronauts weigh a total of 842 pounds.*

(25) *NASA headquarters is in Washington, D.C. NASA's command center is near Clear Lake, Texas, twenty miles from Houston. The primary launch site (Kennedy Space Center) is located on Cape Canaveral, fifty miles from Orlando, Florida.*

(26) *The first American EVA was performed by astronaut Edward H. White II on June 3, 1965, when he left his* Gemini IV *spacecraft and confronted the rigors of space protected only by a spacesuit.*

(27) *The first close fly-by of Mars was made by the Soviet* Mars 1 *mission in 1963. This mission suffered a communications failure and so sent back no useful information.*

(28) *The first Mars landing was made by the Soviet* Mars 6 *mission in 1974.*

(29) Pioneer 4 *was the first successful American fly-by of the Moon.* Pioneer 10 *was the first spacecraft to leave the solar system (after having been the first to pass through the asteroid belt, and the first to investigate the planet Jupiter).*

(30) *The missions' instruments revealed that Jupiter has both auroral displays and lightning.*

(31) *He was a Russian "cosmonaut" (the Russian equivalent of an American "astronaut") Valery Ryumin. His orbital travel in space had totaled 150 million miles.*

(32) *Strangely, it was to conserve fuel.* Ulysses' *trajectory was designed to "slingshot" around Jupiter, by taking advantage of Jupiter's strong gravitational field.*

(33) *The message basically says, "We're here!" It contains basic information about physics, astronomy, and the people of Earth. It is aimed at a galactic cluster called Messier 13. The hope is that some intelligent, technological species in another galaxy may receive the message and know that another technological species exists in the universe.*

(34) *Scientists reason that if life can survive in the Antarctic, it is possible that similar organisms might survive near the poles of Mars, which in many respects are environments similar to that of our coldest landmass.*

(35) *Due to the "time-dilation" effect predicted by Einstein's Special Theory of Relativity, published in 1905. One of the consequences of relativity theory is that a rapidly moving "clock" (the "clock" can be either an ordinary timepiece or a biological organism) moves more slowly the more rapidly it moves through space.*

(36) *A tachyon is a hypothetical particle whose slowest speed is the speed of light.*

(37) *Temporary gains of about two inches in height are common among astronauts who spend extended periods free of gravity.*

(38) *Cosmic rays are highly energetic particles whose origin is still a scientific mystery. They most commonly consist of bare atomic nuclei moving at speeds just below that of light. Because they are an energetic form of "ionizing radiation" they can cause damage to biological cells and to delicate electrical components.*

(39) *Exobiology, the science of organisms that might be found living in environments other than that of Earth.*

(40) *In 1943 Enrico Fermi, the famous physicist, asked "Where are the extraterrestrials?" and this question gave birth to what exobiologists call the Fermi Paradox. Because it seems reasonable to assume that intelligent life exists on other planets, and that some of these speciales should be more*

advanced than humans, this so-called paradox asks, "Why haven't we seen positive evidence of the existence of other intelligent species in the universe?" Scientists at this time have no good answer to this puzzle.

(41) *You know that a UFO is an unidentified flying object. Well, an IFO is an identified flying object. Many reports of UFOs, such as strange lights in the night sky, turn out to be IFOs, such as the planet Venus sinking low in the west.*

(42) *Project Cyclops is a drawing-board proposal designed to aid the search for extraterrestrial intelligence (SETI). It envisions a massive array of antennas designed to pick up incoming radio signals from any others who may be out there in space.*